S0-DZH-059

111 DIGITAL & LINEAR IC PROJECTS

BY DON TUITE

No. 780
$8.95

111 DIGITAL & LINEAR IC PROJECTS

BY DON TUITE

FIRST EDITION

FIRST PRINTING— AUGUST 1975

Printed in the United States
of America

Hardbound Edition: International Standard Book No. 0-8306-5780-0

Paperbound Edition: International Standard Book No. 0-8306-4780-5

Library of Congress Card Number: 75-13013

Cover photo by Micheal J. Pothier

Contents

Preface

What sort of book is this? Primarily, this is a book for experienced electronic hobbyists, technicians, and students who want practical information on how to use linear integrated circuits.

Then, too, this is an idea book. There are well over 100 circuit diagrams for practical projects here. Each identifies all resistance, capacitance, inductance, and supply voltage values needed to make it work. There is enough additional material in the text to permit the reader to make substitutions where substitutions are possible.

This is also a learning book. The reader who devotes sufficient attention will find that he comes away with a better ability to understand digital and linear IC projects he encounters in future reading. He will also be able to design simple projects on his own with confidence that he will be able to successfully troubleshoot them and make them work.

Finally, this is a data book. It provides basic performance data and basing diagrams on almost fifty common and uncommon linear IC types.

The author claims no credit for any but a few of the projects in these pages. Most were created by the talented engineers at National Semiconductor, RCA, Fairchild, Motorola, and Signetics. In a few cases, I made modifications to make a circuit simpler. In a few others, I applied some simple logic to extend the capability of an IC to make it applicable to fields other than intended by its original design.

A word of advice on getting your hands on these linear ICs. Some of the most common are available from national hobby distributors such as Lafayette and Radio Shack. Many more are available only from wholesalers and manufacturers' representatives. Dealing with these sources takes a little patience. If you are serious about getting a certain IC, check your phone book or the phone directory of the nearest large city for electronic parts wholesalers and talk with their sales people. On the West Coast, Almac Stroum and Liberty Electronics have been very cooperative. If you cannot find a wholesaler or manufacturer's rep who will deal with you, write to Allied Radio in Chicago for their industrial catalog ($5.00). They will probably list the item you want in that catalog. If they don't, write them and ask whether they can get it for you. Politeness, persistence, and a willingness to buy a certain minimum order will usually pay off.

And when you do get your parts and you build your project—may it work right the very first time!

Don Tuite

Understanding Linear ICs

In order to use linear integrated circuits effectively, it helps to understand a little of how they are made. The process starts with square sided ingots of pure silicon, approximately an inch and a half on a side. These are sliced into ten-mil-thick wafers which form the substrates of the finished microcircuits. There can be anywhere from 250 to 1000 integrated circuits on each wafer. After the fabrication processes are complete, each circuit is computer checked, defective circuits noted, and the wafer scored and "diced" to separate the individual IC chips. The defective circuits are discarded, and the good ones are mounted in TO-5 type cans, dual in-line packages (DIPs), or flat-packs.

The process of turning a mirror-polished wafer of pure silicon into several hundred integrated circuits begins in an oven heated to 1000 to 1300° Celsius. Here, pure oxygen is blown across the surface of the wafer, causing a thin layer of silicon dioxide to form. The silicon dioxide layer protects the surface of the wafer and provides electrical insulation.

The next step in making the various components that will eventually form the finished microcircuit is the application of a chemical photoresist to the silicon dioxide layer. When this photoresist is selectively exposed to ultraviolet light through a film pattern mask, the exposed portions of the resist polymerize. This renders them insensitive to the developer fluid which is then applied to the wafer. The net result is a pattern of resist on the surface of the silicon dioxide layer that is a negative image of the pattern mask.

The wafer is next subjected to hydrofluoric acid, which etches "windows" through the silicon dioxide layer wherever it is unprotected by resist. When the wafer is placed in a chemical diffusion apparatus like the one in Fig. 1-1, donor (*p*-type) or acceptor (*n*-type) impurities are introduced into the silicon crystal lattice in the region beneath the windows.

Following the initial doping, the silicon dioxide layer is again regrown and the process is repeated to introduce *p*-type

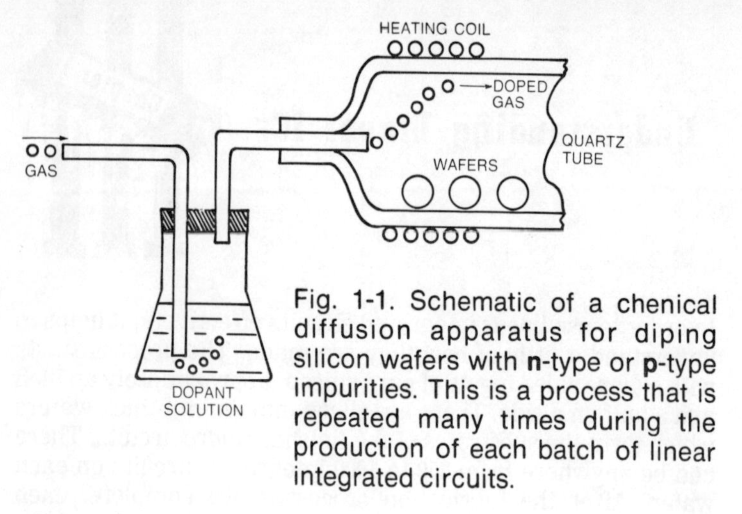

Fig. 1-1. Schematic of a chenical diffusion apparatus for diping silicon wafers with **n**-type or **p**-type impurities. This is a process that is repeated many times during the production of each batch of linear integrated circuits.

regions into *n*-type material and vice versa to produce the required resistors, capacitors, and transistors. As a final step, there is one more etching of the silicon dioxide layer, followed by the vapor deposition of a film of pure aluminum. The aluminum makes contact with the doped regions of the semiconductor through the windows left by this last etching. Then, the aluminum itself is etched to form the pattern of interconnecting leads required to tie together the individual components on each microcircuit.

IC ELEMENTS

There is a difference between the resistors, capacitors, and transistors we use in conventional circuits and those we find on microcircuits. Integrated circuit transistors resemble their discrete counterparts in all ways but one. Figure 1-2 shows a typical monolithic transistor in cross section. The upper portion of the silicon chip, doped with *p*-type donor

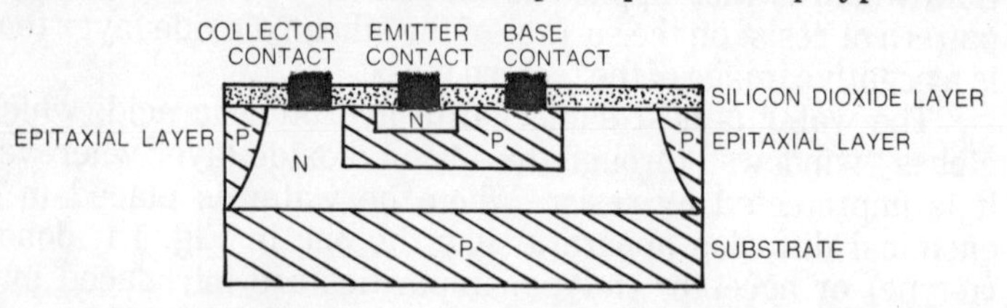

Fig. 1-2. Cross section of a typical monolithic **npn** transistor. To isolate this transistor from other devices in the integrated circuit, the **pn** junction between the collector and the epitaxial layer must be reverse biased.

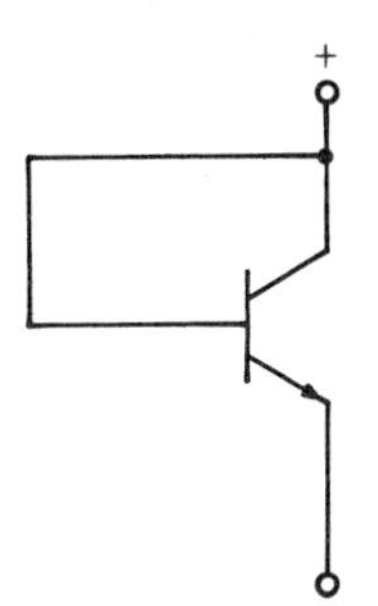

Fig. 1-3. Integrated circuit diodes are fabricated by connecting base and collector of IC transistors. This produces a device with temperature characteristics matched to those of the transistors with which it interacts.

atoms, is called the substrate. On top of this is the *n*-type collector, its boundaries limited by a *p*-type epitaxial layer. Within the collector is a *p*-type base region, and within this is an *n*-type emitter. The important way in which this transistor differs from a discrete transistor is the manner in which it is isolated from other circuit components. Since the *p*-type material in the epitaxial layer is electrically conductive, the *pn* junction between the collector and the epitaxial layer must be reverse-biased to isolate the transistor. In an alternative manufacturing process, deep grooves are etched around the collector down into the least heavily doped part of the substrate, and a silicon dioxide layer is grown on the surface of the grooves, providing isolation. This process allows a denser population of high voltage transistors in a given chip area.

Integrated circuit diodes are rarely made from simple *pn* junctions. Rather, they are forward-biased transistors, as in Fig. 1-3. Their diode thermal characteristics are then essentially the same as the thermal characteristics of the transistors in the circuit. Zener diodes are made the same way, but with a reverse bias, as in Fig. 1-4.

There are two ways of making monolithic integrated circuit resistors. Both involve doping an area of silicon to produce a certain resistivity. The dimensions of this area are selected to give the required resistance. In the simple diffused resistor shown in Fig. 1-5, a layer of *n*-type silicon isolates the

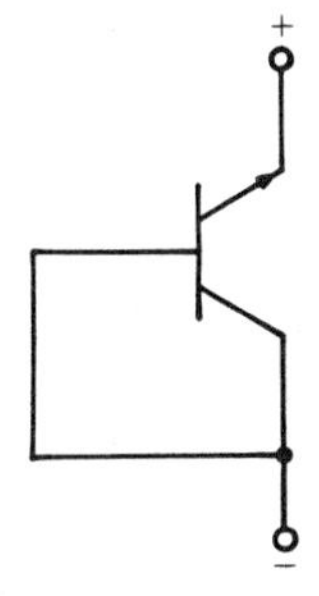

Fig. 1-4. Reverse-biasing the same diode-connected transistor produces a zener diode with a voltage drop of approximately 7V and a negative temperature coefficient. If this zener is connected in series with the forward-biased diode of Fig. 1-3, a zero temperature coefficient zener with a voltage drop of approximately 7.7V is obtained.

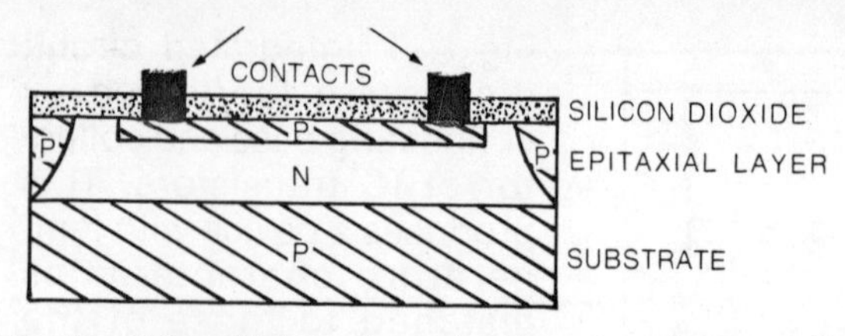

Fig. 1-5. In this diffused resistor, a volume of **p**-type material provides the resistive element.

p-type resistor from the substrate. This type resistor can be produced with resistances from around 100 to 25,000 ohms. In the *pinch* resistor, Fig. 1-6, the *p*-type resistive layer has been cut into by an *n*-type layer doped above it. This constricts the resistive region and produces resistors from 10 to 500K.

Integrated circuit resistors cannot be produced with great precision. A typical value tolerance is ±30%. On an IC chip, however, *ratios* of resistance can be controlled to within 2–3%. Designers therefore use circuits that are dependent on ratios rather than on absolute resistances.

Integrated circuit capacitors can be of two types. For values of a few picofarads, the capacitance associated with the space-charge layer to either side of a reverse-biased *pn* junction is used. Where capacitances up to 30 pF are required, MOS capacitorss like the one in Fig. 1-7 is employed. A highly doped *n*+ region in an *n*-type layer is used as one plate of the capacitor, the silicon dioxide layer is used as a dielectric, and a portion of the final aluminum conducting layer forms the second plate.

ARRAYS AND CIRCUITS

The simplest linear ICs are mere arrays of diodes or transistors. The RCA CA3019, for example, consists of six matched diodes—four in a bridge arrangement, and two isolated. The closely matched diode characteristics of this

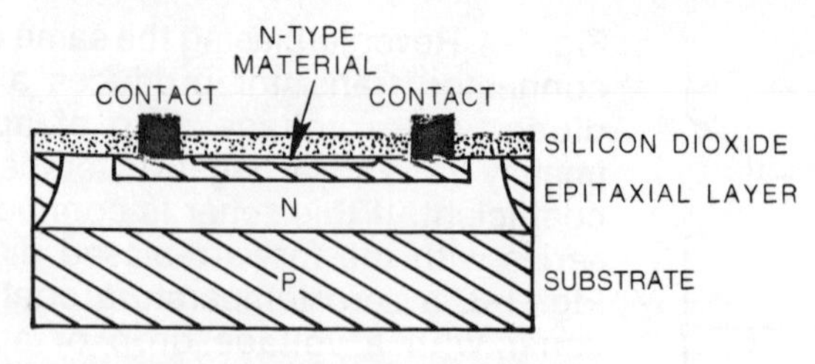

Fig. 1-6. This **pinch** resistor reduces the volume of resistive material by diffusing **n**-type material into it.

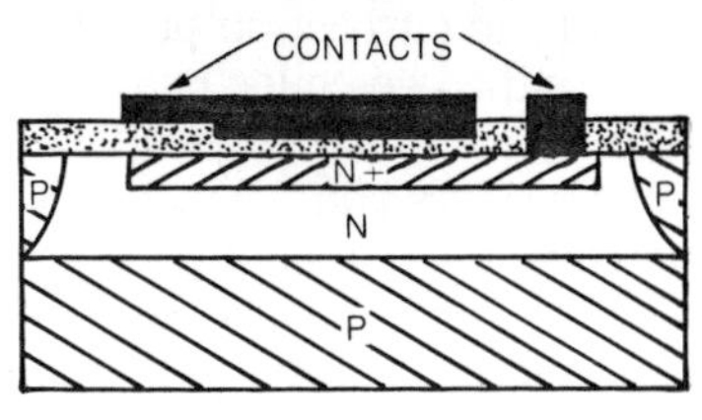

Fig. 1-7. In this MOS capacitor, one plate is formed by an n+ region in the epitaxial layer, while the other is formed by the same aluminum that is used to make contact with the various devices on the chip. The silicon dioxide layer forms the dielectric.

array would make it suitable for a balanced modulator in an SSB exciter. The RCA CA3081 features an array of seven *npn* transistors with emitters connected to a common bus. RCA suggests that among the suitable uses for this array would be that of driving a high-current seven-segment numerical display.

Difference Amplifiers

More complicated linear integrated circuits, including all operational amplifiers, voltage followers, regulators, and functional blocks, are based on one key circuit: the difference amplifier. The circuit in Fig. 1-8 is a prototype difference amplifier. Note that it includes a constant-current generator connected to the emitters of two transistors. We will examine the circuit of this constant-current generator shortly; but first, let's look at how the difference amplifier itself works.

In the case of ideally matched transistors, when no signal is applied to either the inverting or noninverting input, each transistor's emitter current (and, essentially, its collector

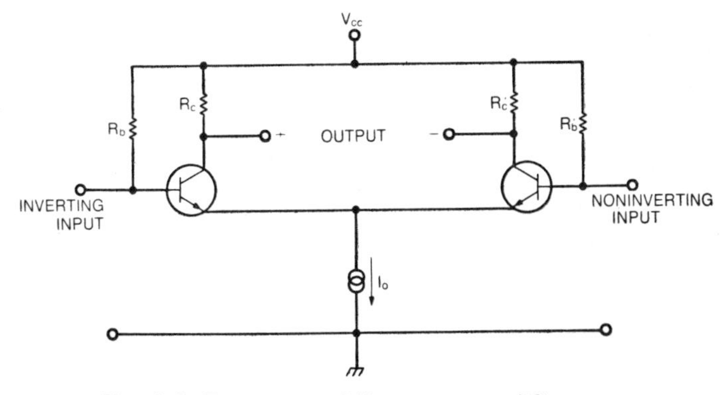

Fig. 1-8. Prototype difference amplifier.

current) is equal to half the current output of the generator. The voltage drops across the two collector resistors are equal and the output of the amplifier is, consequently, zero volts. Naturally, if the two transistors and the two collector resistors are not perfectly matched, the currents through them will not be exactly equal; but because of the proximity of the resistors and transistors on the microcircuit chip, the components are well matched, and there is very little error. The small voltage, usually on the order of a few millivolts, that must be applied between the inverting and noninverting inputs of a practical microamplifier in order to obtain zero output is called the offset voltage of the amplifier.

Common-Mode Signals

Suppose one signal is applied simultaneously to the inverting and noninverting inputs of a difference amplifier. What happens? Since the balanced nature of the circuit is not disturbed by this signal, the output voltage of the amplifier is still zero. An input signal with the same frequency, phase, and amplitude at both the inverting and noninverting inputs is called a *common-mode* signal.

Differential-Mode Signals

Then what happens when a signal is applied so that its polarity at the inverting input is opposite to its polarity at the noninverting input? In this case, the currents through the two transistors are unequal, and an amplified version of the input waveform appears across the output terminals. Input signals like this are called *differential-mode* signals.

In a typical application, unwanted noise, power-line hum, clicks, transients, etc. appear at the input of a difference amplifier as common-mode signals; that is, they show up equally on both signal leads. A microcircuit having a difference amplifier as its input stage ignores these signals and amplifies only the desired differential-mode signal.

Common-Mode Gain and Rejection Ratio

Actually, since practical microamplifiers are not perfect, some of the common-mode signal will appear at the output. One of the characteristics that manufacturers of microcircuits list on their specification sheets is common-mode gain. Since this is typically on the order of −30 dB, you can see it is actually quite a loss. Another figure of merit is the *common-mode rejection ratio*, which is the algebraic difference between the differential-mode gain and the common-mode gain. Since many microcircuit amplifiers have

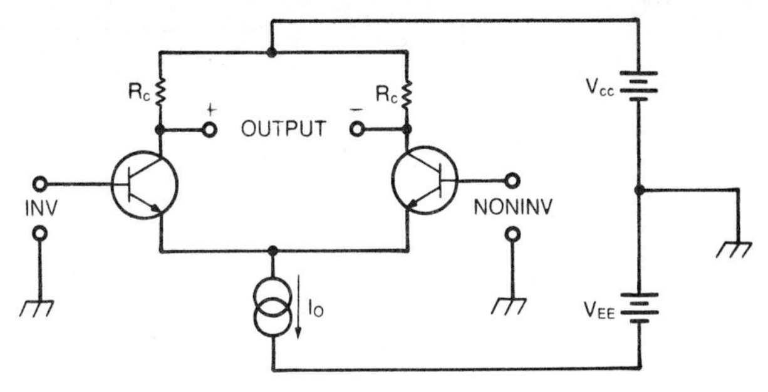

Fig. 1-9. Practical difference amplifiers for integrated circuits must use external biasing. This is the reason that most linear integrated circuits require dual-voltage supplies.

differential-mode gains of 60 dB and more, a typical figure for common-mode rejection ratio is on the order of 90 dB.

Practical Difference Amplifiers

The basic difference amplifier of Fig. 1-8 cannot as yet be manufactured as a practical microcircuit. This is because the bias resistors, R_b and R_b' would have to be on the order of several hundred kilohms or more. Presently, this is just too much resistance for monolithic circuit fabrication techniques. In practice, therefore, the circuit in Fig. 1-9 is used. This is simply the same circuit with an external bias supply. Most microcircuit amplifiers are designed to operate from dual voltage supplies, although there are ways of using single-voltage supplies (which we discuss in a later section).

Many microcircuit amplifiers use a darlington-connected pair instead of a single transistor in each side of their input

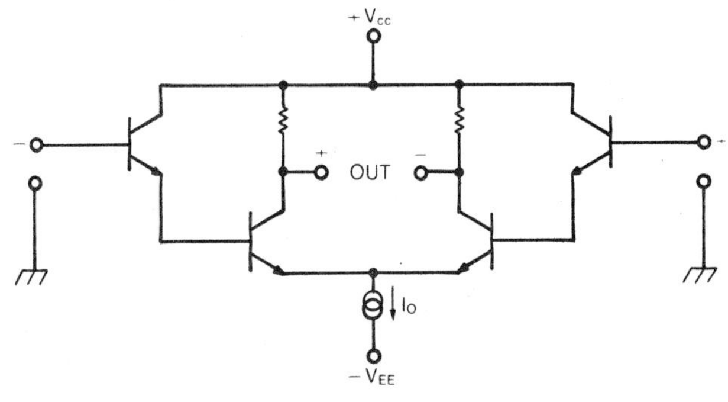

Fig. 1-10. A difference amplifier with darlington-connected transistors.

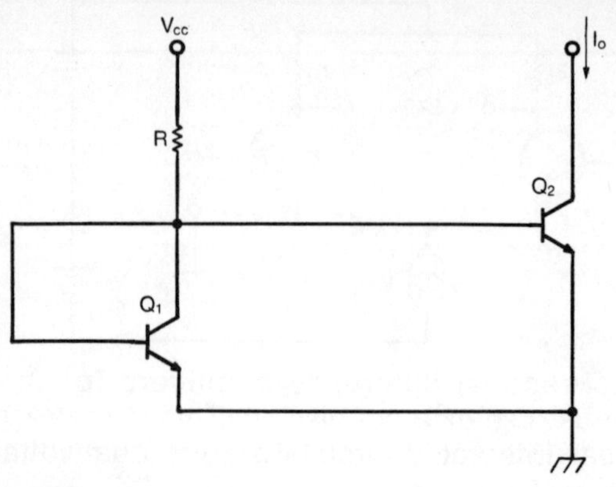

Fig. 1-11. Typical current generator for integrated-circuit difference amplifiers.

difference amplifier, as shown in Fig. 1-10. This provides a higher output with less loading of the signal source.

The constant-current generator shown in the difference amplifiers can be realized using the circuit of Fig. 1-11. Transistor Q1 tends to keep the operating point of Q2 fairly constant in the face of temperature and power supply fluctuations, and this in turn keeps the collector current of Q2 constant.

The difference amplifier we have been considering exhibits what is called a *double-ended* output. That is, the output voltage is not referred to ground, but floats instead at a potential other than ground voltage. The difference microamplifier in Fig. 1-12 is used in many integrated circuits to change the double-ended output of a previous difference

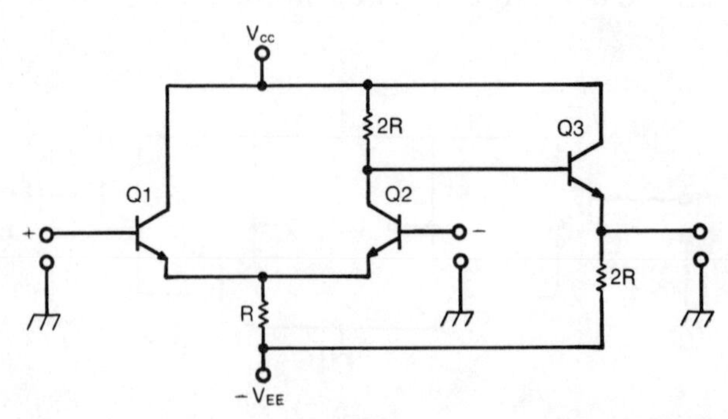

Fig. 1-12. A difference amplifier with a single-ended output. Transistor Q3 is used as an emitter-follower impedance transformer.

amplifier to a *single-ended* output — that is, one in which a differential mode input signal of zero volts produces an output signal zero volts above ground. A similar difference amplifier using darlington-connected transistor pairs instead of single transistors is found in many microamplifiers.

OPERATIONAL AMPLIFIERS

Operational amplifiers or *op-amps* form the most ubiquitous type of linear integrated circuit. Most feature a difference amplifier input stage, another difference amplifier buffer and level shifter, and frequently, a class B output stage. The operational amplifier is characterized by three features: very high input impedance, very low output impedance, and extremely high gain. For a typical operational amplifier, the 741, these characteristics are: input impedance—1M, output impedance—75Ω, differential-mode voltage gain—15,000 (approximately 84 dB). A word of explanation here about certain common op-amp types, notably the 709 and 741. The original 741 was the Fairchild μA741. Now, all manufacturers make similar devices with 741 somewhere in their nomenclature—e. g. RCA's CA741C, National Semiconductor's LM741, etc. All of these types are directly interchangeable in terms of performance and pin assignments.

Negative Feedback in Op-Amps

In circuit diagrams, operational amplifiers are represented by triangles, as in Fig. 1-13. The inverting and noninverting inputs are designated by minuses and pluses, respectively, along one side; and the output, in most cases, is taken as the point of the triangle opposite this side. The utility of the operational amplifier comes about as a result of its behavior when a negative feedback path is supplied. Consider the circuit in Fig. 1-14. We will assume that the source voltage, V_S, is small enough that it does not quite drive the amplifier to saturation. If the input impedance of the amplifier is very

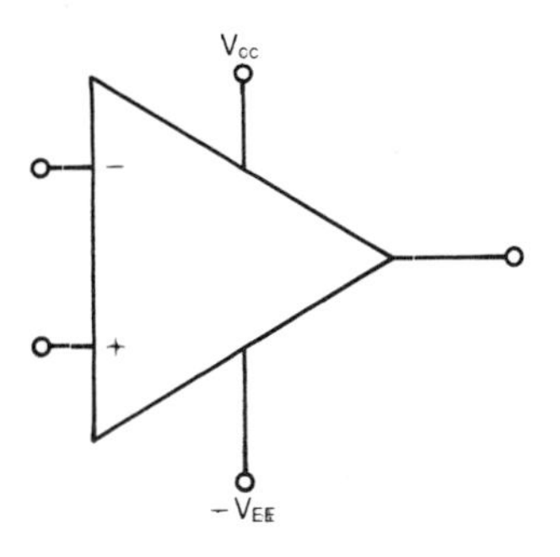

Fig. 1-13. Basic circuit symbol for an operational amplifier. Power connections are frequently omitted to simplify drawings. Other terminals may be drawn along the sides of the symbol to represent connections for compensation networks.

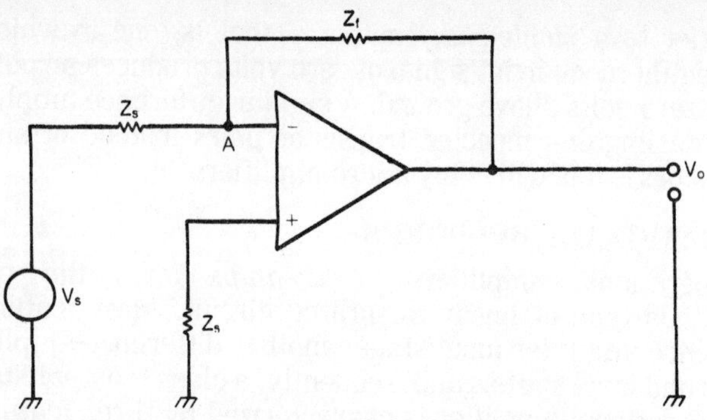

Fig. 1-14. The loop gain of the operational amplifier with nevative feedback is dependent solely on the ratio of feedback impedance to source impedance.

large, we may consider the current flowing into the inverting input of the op-amp to be zero. If the gain of the op-amp is very large, the voltage at the inverting input must be quite small. For instance, if this a 741 op-amp, with a gain of 15,000 for a 15V output signal, the input can only be 1 mV *at the inverting input*. This is not to say that V_s is only 1 mV, as we shall see. If we consider this small input voltage to be zero, the analysis of the circuit in the figure becomes very simple.

The current from generator V_s toward point *A* must be simply V_S/Z_S. Likewise, the current through the feedback impedance away from point *A* must be simply V_o/Z_{fb}. If there are no other currents into or out of this point, then these two must be equal, and

$$\frac{V_S}{Z_S} = \frac{V_O}{Z_{fb}}$$

We can rearrange this to find V_o/V_s, the closed-loop voltage gain:

$$\frac{V_O}{V_S} = \frac{Z_{fb}}{Z_S}$$

In other words, the gain of the circuit is solely dependent on the feedback impedances. If Z_S and Z_{fb} are both one megohm resistors, the circuit gain is unity. If Z_{fb} is 1M and Z_S is 10K, the circuit gain is 100.

Impedances Z_S and Z_{fb} need not be purely resistive. If a capacitor is connected from the op-amp output to its inverting input and a resistor is used in series with the input, as in Fig. 1-15, the circuit functions as a mathematical integrator. This is discussed in more detail in a later chapter.

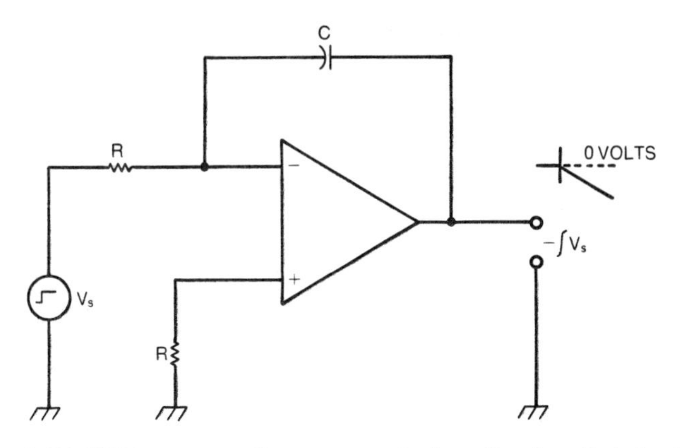

Fig. 1-15. With a capacitor connected as the feedback element, an operational amplifier mathematically integrates whatever function is applied to its input. In the figure, the input is a step function and the output is a ramp. Note that the ramp is inverted, indicating a sign change between input and output.

If we use even more elaborate combinations of resistors and capacitors in the feedback and input circuits, we can create high- and low-pass, bandpass, and band-elimination filters (Chapter 7).

Note that we have been connecting the noninverting op-amp terminal to ground through a resistor in these feedback circuits and supplying our signal to the inverting input. This produces a 180° phase shift between output and input. We can connect our signal to the noninverting input if we modify the circuit as in Fig. 1-16 to eliminate the phase shift. However, the expression for circuit gain is no longer what it was. If you understand small-signal models, Fig. 1-17A will help explain. The portion of the circuit enclosed in the dashed lines represents the op-amp. It consists of two input terminals, V1 and V2, with infinite impedance to ground, and a voltage source with an output, V_o, equal to a constant, k, times $(V2 - V1)$. Given the external connections of Fig. 1-16, we note that the output voltage is equal to $V_o - V_s$. Consequently, the current through Z_{fb} toward V_1 is $(V_o - V_s)/Z_{fb}$, and the current through Z_s away from V_1 is V_s/Z_s. Assuming the currents to be equal and proceeding as before to find the closed-loop gain, V_o/V_s, we find

$$\frac{V_o}{V_s} = 1 + \frac{Z_f}{Z_s}$$

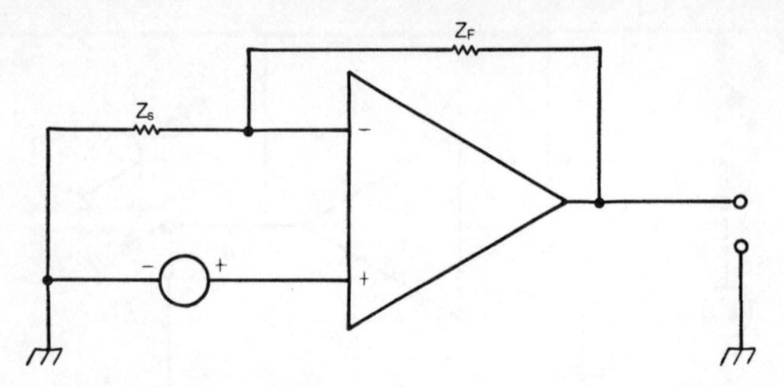

Fig. 1-16. An operational amplifier connected for zero phase shift between input and output. In this case, the loop gain is equal to one plus the ratio of feedback to source impedance.

In cases of large gain and purely resistive impedances, the one on the right side of the equation can be neglected. However, in cases where Z_f or Z_s is reactive, or even where Z_f and Z_s are purely resistive, but $Z_f \leq Z_s$, the one becomes significant.

Voltage Followers

A special case of an operational amplifier used in the noninverting mode is the voltage follower. In the voltage

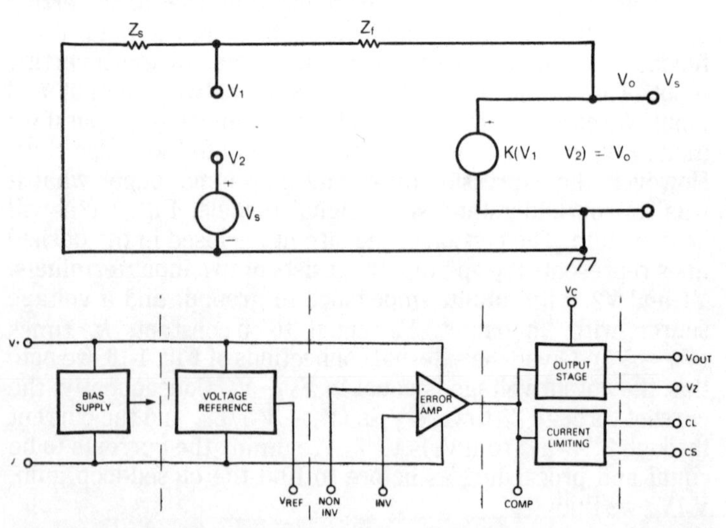

Fig. 1-17. (A). Small-signal model of noninverting op-amp circuit shown in Fig. 1-16. (B). Block diagram of a 723-type voltage regulator.

follower, Z_{fb} is a short circuit (100% feedback), and a fairly large impedance is used in the input. Voltage followers are used to transform high impedances to low. Of course, their gain is essentially unity.

Compensation

Some types of operational amplifiers require external components connected between the collectors of their first-stage difference amplifier transistors to prevent self-oscillation. The 709 is an example of this. Pins *1* and *8* of the TO-5 version of the 709 should be connected by a 100 pF capacitor and a 1.5K resistor in series in the external circuit. Other terminals are provided in order to provide uniform gain over high bandwidths. To extend the bandwidth of the 709 to approximately 2 MHz, a 3 pF capacitor should be connected between the output (pin *6*) and pin 5. Specific circuits in later chapters illustrate more of these techniques.

VOLTAGE REGULATORS

Add a voltage reference and an output stage to an operational amplifier and you have the basis for a voltage regulator (Fig. 1-17B). Practical regulators usually add a current limiter to protect the device from short circuits across the output.

A Stable Voltage Reference

The voltage reference portion of a voltage regulator provides a potential that remains constant in spite of large variations in supply voltage and temperature. It is interesting to see how linear integrated circuit designers achieve this voltage stability. Figure 1-18 shows a typical circuit. The zener

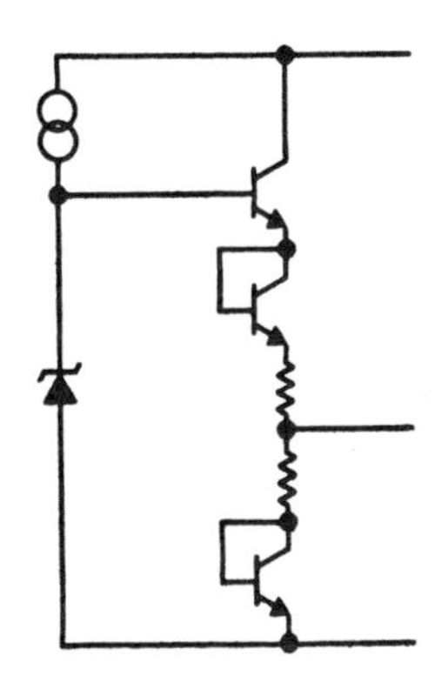

Fig. 1-18. Voltage reference section of a typical voltage regulator IC.

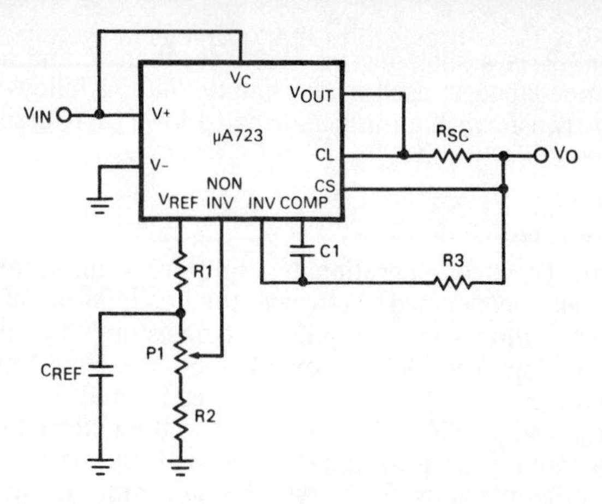

Fig. 1-19. A linear IC voltage regulator circuit. Typical performance: output voltage, 5V; line regulation, 0.5 mV for ± 3V line voltage variation; load regulation, 1.5 mV for ± 50 mA variation in load current.

diode, D1, is supplied by a constant current source as described earlier. This prevents power-supply fluctuations from affecting the reference voltage. The voltage provided by the zener has a positive temperature coefficient of around 2.4 mV/°C. To compensate for this, the zener output is applied to an emitter follower and voltage divider in series with two diode-connected transistors. The voltage divider reduces the positive temperature coefficient of the reference voltage to a level that is exactly balanced by the negative temperature coefficients of the transistors, so that there is no net voltage change for any change in temperature.

Figure 1-19 shows the operation of a typical voltage regulator. The reference voltage is applied across a resistive divider, and a portion of it—determined by the ratio of the resistances in the divider—is applied to the noninverting input of the device's op-amp. The output voltage of the regulator is fed to the inverting input of the op-amp. The op-amp drives the output stage to the point where the voltages at the inverting and noninverting stages are equal and adjusts the output stage to maintain this voltage in spite of load or supply variations.

Some additional external components are necessary in practical circuits. A small capacitance, on the order of 100 pF, is needed between the output of the op-amp and its inverting input to keep the device from going into oscillation. Resistor R3, equal to the parallel equivalent of voltage-divider resistors R1 and R2, minimizes temperature effects caused by R1 and

R2. Finally, R_{SC} is provided to limit current in the event the output is short-circuited.

Higher Current Regulators

Most regulator ICs can't handle more than about 100–150 mA of output current, but it is relatively simple to add one or more power transistors outboard of the regulator to handle levels of up to about 2A. After a point, determined by the dissipation requirements for the transistors available, it may become easier to change over to a switching regulator arrangement, as illustrated in Fig. 1-20. In the switching regulator, transistor Q1 is biased either full on or full off by the output of the regulator IC, and the supply voltage is fed through Q1 and an LC filter to the load. The power dissipation requirement of Q1 is determined by the duty cycle of the pulses it carries. The voltage at the load is, as in the conventional regulator, fed back to the inverting input of the regulator's op-amp. The purpose of the filter is to smooth the pulses from Q1 into low-ripple DC. Diode D1 protects the regulator and Q1 from the back-voltage spikes produced by the coil each time Q1 shuts off. Of necessity, this cannot be a conventional silicon diode, but must be a special, very fast-acting device.

LINEAR INTEGRATED CIRCUITS FOR RF

The present state of linear IC art limits usable frequencies to those below about 50 MHz. In this portion of the RF spectrum, linear ICs find wide use as video and IF amplifiers. Figure 1-21 shows an IF amplifier/AM detector with built-in AGC.

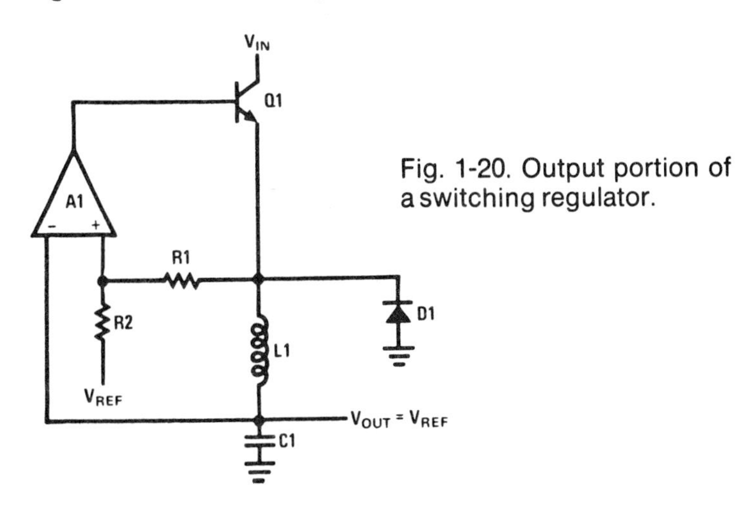

Fig. 1-20. Output portion of a switching regulator.

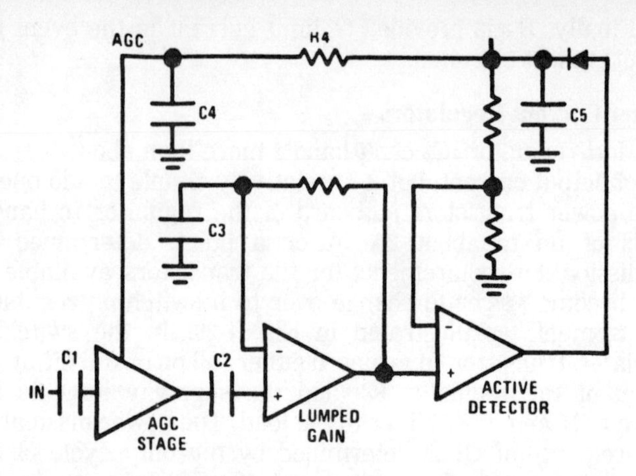

Fig. 1-21. Block diagram of a linear IC IF amp/AM detector with built-in AGC(National Semiconductor LM172).

This is the National Semiconductor LM172. It employs three stages. The first is a variable attenuator controlled by the AGC feedback voltage. This is followed by an extremely high gain stage. The last stage is a sensitive AM detector. No IF transformers are required.

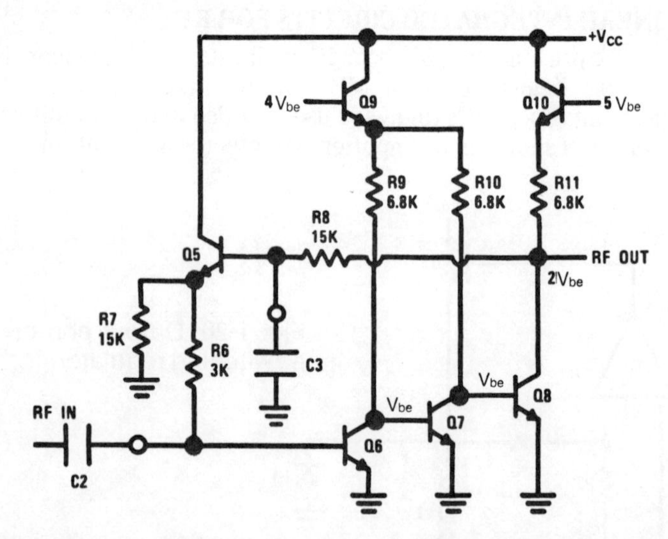

Fig. 1-22. AGC stage of the LM172. Operation is explained in the text. The voltage $3V_{be}$ applied to the base of Q1 is obtained from the voltage drop across three series-connected IC diodes.

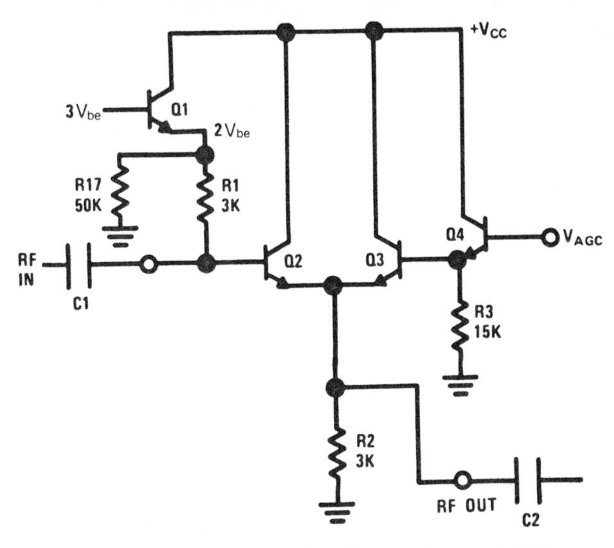

Fig. 1-23. Gain stage of the LM172. Various V_{be} bias values are obtained from voltage drops at various places along a diode string. V_{be} is approximately 0.7V.

It is worthwhile to look at the three stages of this IF amp/detector to see how it differs from a conventional discrete component IF strip. The first stage, a variable attenuator (Fig. 1-22), consists essentially of a pair of transistors connected in a fashion that resembles a difference amplifier. Note, however, that the output is taken across the common emitter resistor of the pair. The IF signal is fed into Q2, which is biased at a fixed level. The AGC signal is fed into Q3. As long as the AGC level is below the bias level of Q2 , Q3 is cut off and Q2 functions as an ordinary emitter follower. When the AGC level to Q3 exceeds the bias level on Q2, Q3 conducts and tends to turn Q2 off. The greater the AGC level, the greater the attenuation.

The second stage (Fig. 1-23) cascades three transistors connected in common-emitter configuration. The other components bias the transistors for class A operation.

Rather than use a single diode as an AM detector, the LM172 uses a diode in the negative feedback loop of a moderate-gain difference amplifier (Fig. 1-24). The amplifier biases the diode, allowing it to respond to smaller signals than it would be able to if it were unbiased. In addition, the amplifier provides an additional voltage gain. The result is an 800 mV (peak-to-peak) audio output when an input signal with 80% modulation is applied.

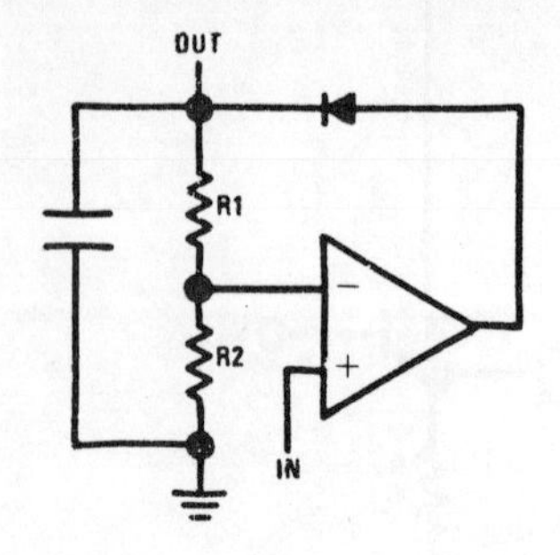

Fig. 1-24. Output stage and detector of the LM172.

PHASE-LOCKED LOOPS

Perhaps the single most exciting thing to come out of linear IC technology is the application of phase-locked loop (PLL) devices to consumer products and hobbyist projects. Before ICs, PLL was limited to a very few military communications systems and super-esoteric amateur projects. Now that the circuit complexity required for PLL can be reduced to a few or even a single integrated circuit, PLL is becoming widely available.

What's a phase-locked loop? Figure 1-25 is a PLL block diagram. Basically, there are three components—a phase detector, a filter, and a voltage-controlled oscillator. In operation, two signals are fed into the phase detector. One is an external signal; the other is the output from the voltage-controlled oscillator. The phase detector produces a DC error signal that is proportional to the phase difference between the outside signal and the signal from the oscillator. This error signal is filtered and applied to the voltage-controlled oscillator to change its frequency to match the frequency of the outside signal.

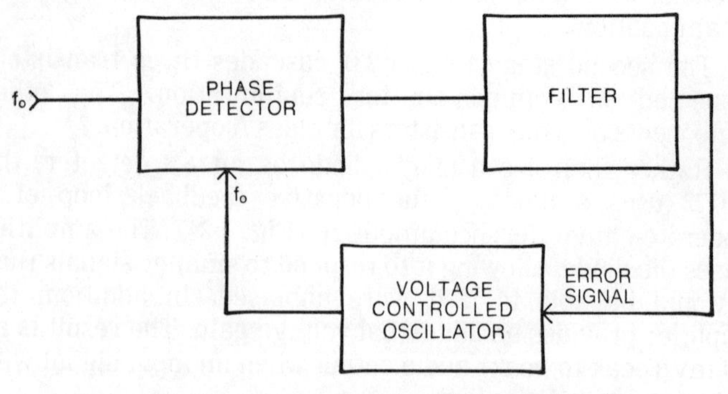

Fig. 1-25. Components of a phase-locked loop.

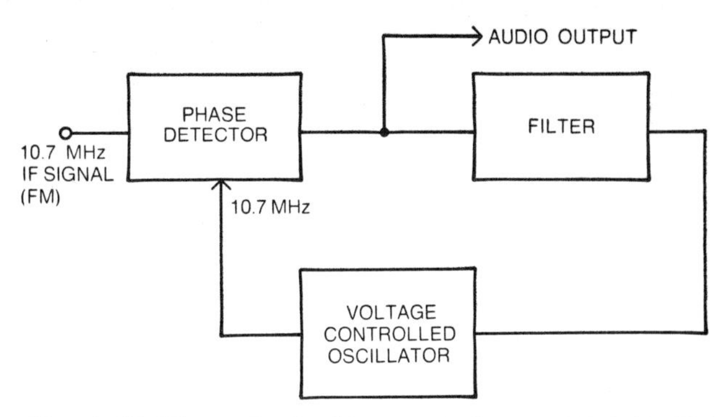

Fig. 1-26. Phase-locked loop used as an FM detector.

PLL FM detectors

You say that doesn't sound so impressive? You're wondering what all the fuss is about? Let's look at two applications of PLLs to see where its value lies. In Fig. 1-26, the outside signal being fed into the phase detector is the IF signal from the front end of an FM receiver. The voltage-controlled oscillator is locked to the IF, the frequency-modulated sidebands having been removed by the filter. The oscillator follows the frequency of the carrier regardless of instability of either the broadcast signal or the receiver front end's local oscillator. Meanwhile, it happens that the phase detector output ahead of the filter corresponds to the modulating signal that was impressed on the carrier. The PLL thus functions as a sensitive, totally drift-free detector for FM and FSK signals.

PLL Frequency Synthesizers

A further application of the phase-locked loop is in frequency synthesis. In Fig. 1-27, a programmable divider has

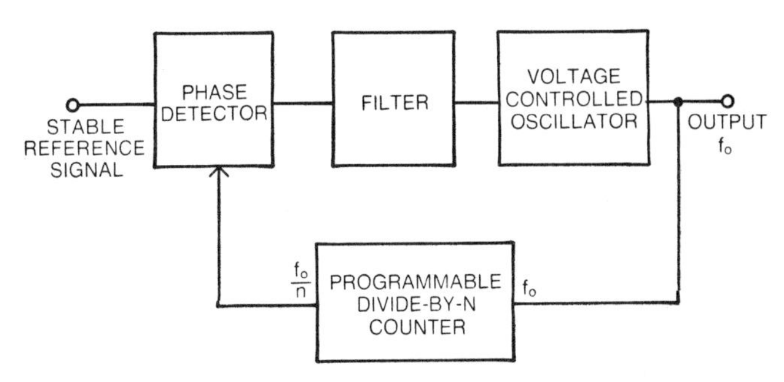

Fig. 1-27. Phase-locked loop used as a frequency synthesizer.

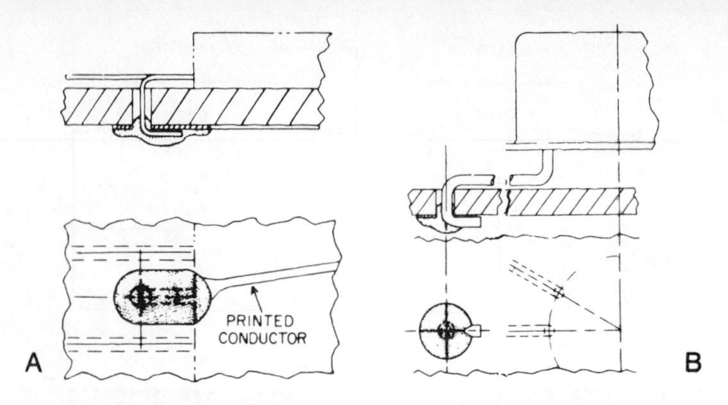

Fig. 1-28. Soldering flat-pack (A) and TO-5 (B) ICs to printed circuit boards.

been added between the output of the voltage-controlled oscillator and the input of the phase detector. In a practical circuit this counter might be externally programed (say, by pressing button switches) to divide by any number from 30 to 40. With the outside frequency set by a very stable crystal-controlled 100 kHz generator and the center frequency of the voltage-controlled oscillator programed for about 3.5 MHz, the oscillator output can be stepped in 100 kHz increments from 3 to 4 MHz by changing the modulus of the frequency divider. Each output frequency will be as stable (on a percentage basis) as the 100 kHz crystal-controlled input frequency, and the precision of the 100 kHz step frequencies will also be consistent with the reference frequency. Obviously, this is many times more stable and precise than any conventional variable frequency oscillator. Moreover, it is compact, uses only one crystal, requires no mechanical switching of an rf signal and, thanks to IC technology, it is relatively inexpensive.

USING LINEAR ICS

Linear ICs are rugged and require no special handling techniques. Naturally, you should avoid dropping your ICs onto hard surfaces or subjecting them to elevated temperatures for long periods of time. If you are soldering directly to IC leads, use a low-wattage soldering iron with a clean, tinned tip.

Installation

For RF applications, it is better to solder the device directly into the circuit rather than use a socket. This will

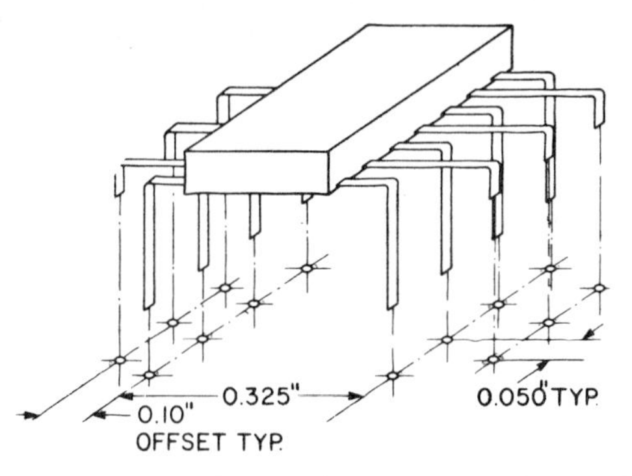

Fig. 1-29. Flat-pack leads can be staggered, as shown, to provide more spacing between adjacent leads.

minimize lead inductance and RF leakage problems. Also, when you are working with RF, it is a good idea to swab your printed circuit board with a liberal amount of alcohol after you finish soldering. The alcohol will dissolve and wash away the solder flux that was deposited on the board during soldering. Flux isn't particularly a problem at low frequencies, but its resistance decreases inversely with frequency, and its presence could lead to RF feedback.

Figure 1-28 shows how to solder flat-pack and TO-5 ICs. As in all electronic soldering, the idea is to form a sturdy mechanical joint first, and then solder. The solder is only to provide electrical continuity; strength comes from the joint.

Staggering Flat-Pack Leads

Particularly with flat-packs, you may find it difficult keeping adjacent leads separated. Figure 1-29 shows how to stagger flat-pack leads to give yourself more room. As Fig. 1-30 shows, the greater spacing between leads in dual in-line packages (DIPs) makes staggering unnecessary. Be very careful when bending IC leads. A nick on a lead creates a stress concentration that can cause fatigue failure. The same is true for very sharp-radius bends. Be careful not to bear against the side or bottom of the IC package with your pliers when you are making a bend in a lead. You may pull a connection loose inside the device.

Guarding

Many high-gain op-amp ICs provide for *guarding* the input pads on the printed circuit board from leakage currents. They do this by leaving the leads on either side of the input leads unconnected to any part of the integrated circuit. This allows

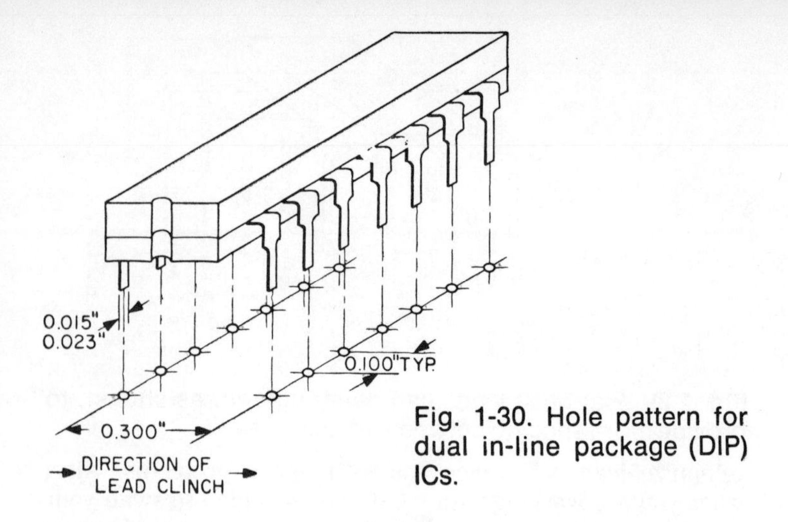

Fig. 1-30. Hole pattern for dual in-line package (DIP) ICs.

the input terminals to be completely surrounded by a ring of conducting material, as in Fig. 1-31. Figure 1-32 shows different ways of connecting the guard ring, either to ground in an inverting op-amp, or bootrapped to the output in a voltage follower.

Dual-Voltage Supplies

As noted in the section on difference amplifiers, these microamplifiers generally cannot be made self-biasing, so dual-voltage supplies are required for most linear ICs. Usually, the requirement is for symmetrical voltages, somewhere in the range from 5 to 15V. Some linear ICs have supply requirements up to ±50V. It is very important that the power connections be bypassed *at the IC* by 0.01 μF or larger capacitors. In cases where it may be possible to connect a supply with reverse polarity, it is a good idea to put series diodes in the supply leads to protect your ICs.

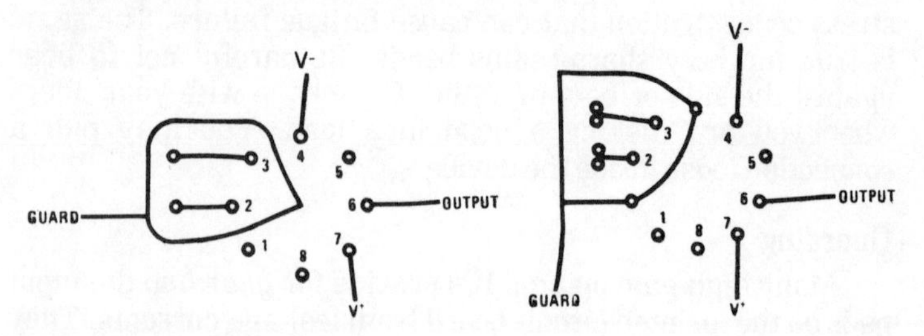

Fig. 1-31. Guarding patterns for 8-lead (A) and 10-lead (B) TO-5 can IC op-amps.

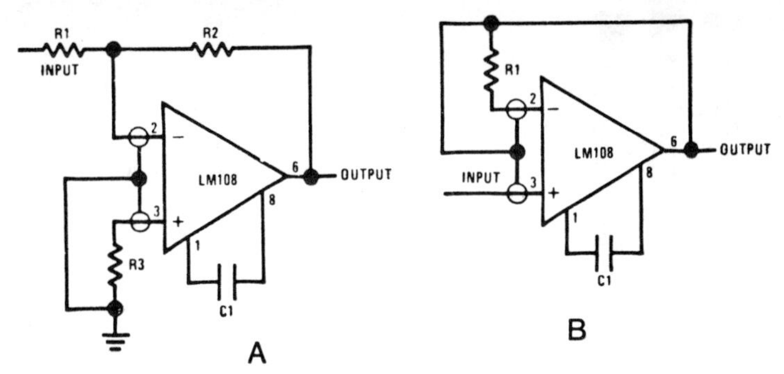

Fig. 1-32. Guard connections for inverting op-amp (A) and voltage follower (B).

Single-Voltage Supplies

In many cases, it is possible to use a single-voltage rather than a dual-voltage supply. It is simply a matter of splitting the supply voltage with a pair of zener diodes, as in Fig. 1-33. Resistor R5 limits the current to the zeners and should be selected to drop the difference between the dual zener voltage and the supply voltage. Capacitors C3 and C4 bypass the power leads; they should be at least 0.01 μF, and they should be installed right at the IC power leads.

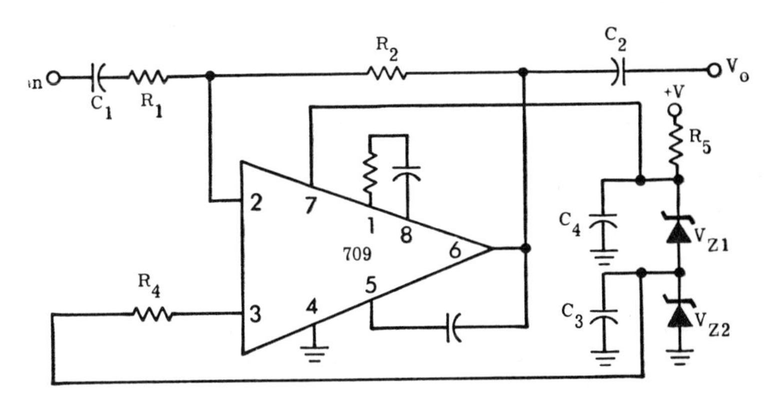

Fig. 1-33. Single-supply operation of a 709-type op-amp.

Audio Amplifiers

Of all the uses to which linear ICs can be put, audio amplification is the one that springs first to mind. In this chapter, we will take a close look at some different kinds of linear ICs and at some audio amplifier projects that incorporate them.

The general form of all the audio amplifier projects is that of an op-amp with negative feedback operated in the noninverting mode. From Chapter 1, recall that the gain of such a circuit is equal to one plus the ratio of feedback to source impedance, so all these amplifier configurations have a minimum of unity gain. From this theoretical starting point, there can be a great deal of diversity in practical circuits, as we shall see.

THE LH0021/LH0041

The National Semiconductor LH0021/LH0041 (Fig. 2-1) is an operational amplifier of more or less classic design. It is, however, capable of supplying fairly large output currents, and this allows these op-amps to drive small loudspeakers with considerable power.

The maximum supply voltage for which these linear ICs is ±18V. Under no-load conditions, they will draw 2 mA from a dual 15V supply at room temperature.

The LH0021 is the higher-power member of the pair. It is capable of a maximum power dissipation of 22W with a suitably large heatsink. To aid in heatsinking, the LH0021 comes in an eight-lead TO-3 case. The LH0041 has a maximum dissipation of only 2.4W. This comes about not only because of its lower current capacity, but because of the difficulty of adequately heatsinking its 12-pin TO-8 package.

Referring to Fig. 2-1, we find a difference input stage with a single-ended output. In the LH0041 only, the terminals marked OFFSET NULL are available for connection of a 10K pot. The output of the difference stage is fed to complementary push-pull emitter followers, the necessary phase inversion for one side taking place in darlington pair Q9—Q10. This use of an active phase-inverter is a good example of the difference between conventional discrete-component design and integrated circuit design. The traditional means of obtaining a 180° phase difference between the inputs to a push-pull stage is a transformer. On an IC, however, not only is a transformer out of the question, but the cost of providing the additional transistors to perform the inverting function is so trivial that an active phase inverter is the *only* way to go.

Notice the two terminals on the output side of the circuit labeled I_{SC}. An external resistor is to be connected from each of these terminals to its associated power input terminal. The size of the resistor determines the amount of current that can flow when the output is short-circuited. Practical values for these resistors range from 0 to 3.3Ω. The larger resistors limit short-circuit current to a smaller value, but they also cut into the power output of the amplifier.

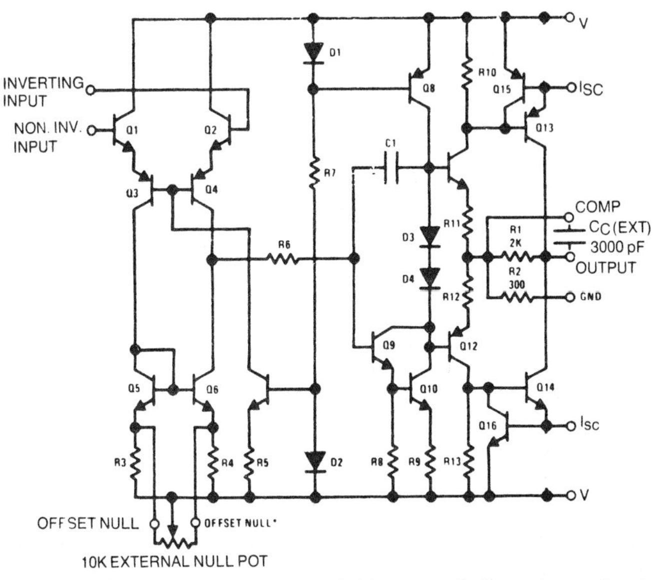

Fig. 2-1. Circuit diagram of National Semiconductor LH0021 and LH0041 operational amplifiers. The LH0021 is capable of power output of 1A at 12V. It is supplied in an 8-pin TO-3 (power transistor style) package with the package case used as the output terminal. The LH0041 is capable of a power output of 200 mA at the same voltage level as the LH0021. It is supplied in a 12-pin TO-8 package. The connections for an external offset null pot shown in the figure are available on the LH0041.

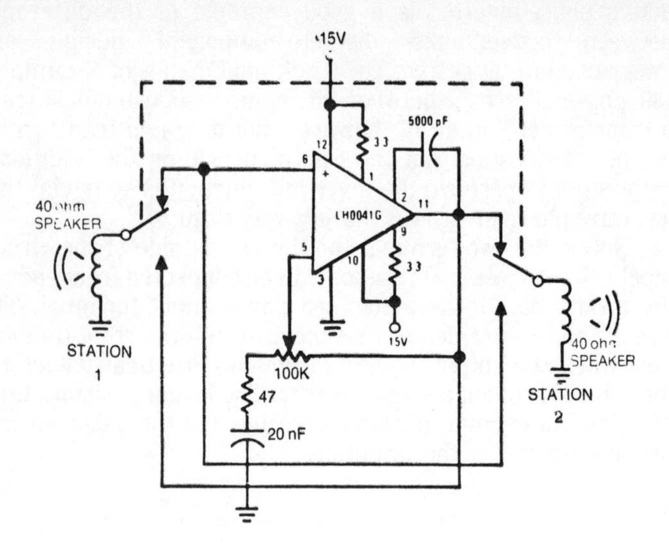

Fig. 2-2. A simple intercom system using the LH0041.

A Simple Intercom

The two-station intercom in Fig. 2-2 illustrates a very simple application of the LH0041. Pins *10* and *12* are the power connections, and pins *1* and *9* are the I_{sc} connections described in the last paragraph. A 5000 pF capacitor is connected from the output to compensation pin 2.

The circuit configuration is the basic noninverting operational amplifier design we discussed in Chapter 1. Part of the output is fed back to the inverting input through part of the 100K potentiometer, while the 47Ω resistor plus the rest of the pot serves as the source resistance. Thus, the gain varies from slightly more than unity to about 90 dB. Since this is the first project in this book, this is a good place to caution again that it is absolutely necessary, in all these projects, to bypass both power supply leads (±15V) with 0.01 or larger capacitors as close as possible to the IC itself.

Combining ICs for More Power

By using two LH0021s in a bridge arrangement, as in Fig. 2-3, we can drive an 8Ω speaker with up to 35W rms. Since one op-amp is fed in its inverting input and the other is fed with the same signal in its noninverting input, a single-ended signal is all that is required.

There's a surprise in the feedback arrangements of the two op-amps—they're not the same! The upper amplifier,

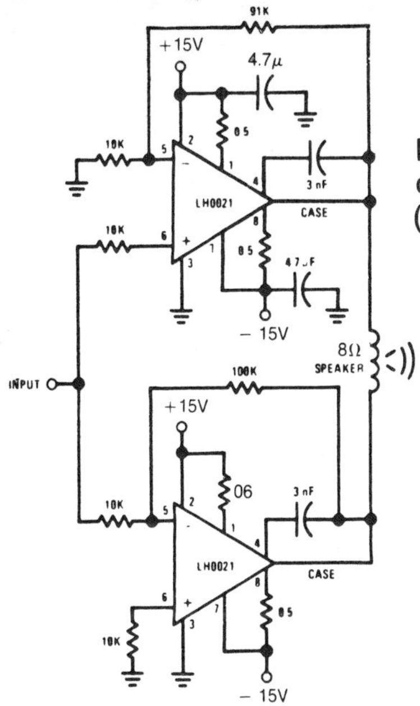

Fig. 2-3. Using two LH0021 op-amps to deliver 35W (rms) of audio power.

which is used in its noninverting mode, has a 91K feedback resistor and a 10K source resistor. This results in a stage gain of 1 + 9.1, or 10.1. The lower amplifier, which is used in its inverting mode, has a 100K feedback resistor and a 10K source resistor. This gives *it* a stage gain of 10. By selecting the feedback components for the two amplifiers in this way, the two stages have been gain-balanced to within 1% of each other.

In the circuit, 4.7 μF bypass capacitors are shown only on the power inputs of the upper amplifier. This can be realized on a practical circuit only if the two ICs are mounted in very close proximity. The 0.5Ω resistors are for short-circuit current limiting and the 3 nF (3000 pF) capacitors between output and pin 4 provide compensation.

THE CA3048 QUADRAPHONIC AC AMP

The very complex-appearing circuit in Fig. 2-4 represents RCA's CA3048. Actually, the CA3048 turns out to be a very simple device, once we see that it consists of one AC amplifier replicated four times. The basic amp, it turns out, is even simpler than most, since it requires no external compensation and is designed to operate from a single 9–16V supply. Nominal supply voltage for most applications is 12V.

Let's look at a representative amplifier, the one in the upper left corner of the diagram. Pins 7 and 8 are the inverting

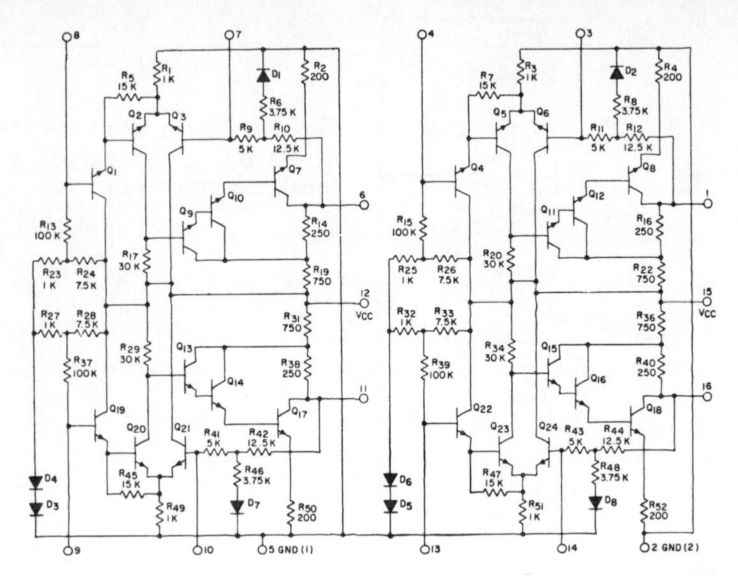

Fig. 2-4. At first glance, this circuit diagram for the RCA CA3048 quad op-amp appears intimidating. Close inspection, however, reveals four identical amplifiers of simple design. Each comprises a level-shifting difference amplifier input stage and a darlington pair feeding a common-emitter output stage.

and noninverting inputs, respectively, of a difference amplifier consisting of Q2 and Q3. The single-ended output of this amplifier stage is fed, via a darlington pair consisting of Q9 and Q10, to output transistor Q7. The output, taken from the collector of Q7, is available at pin 8. An internal resistive network provides feedback from the output to the inverting input, allowing gain to be controlled simply by varying an external input resistor on the inverting side.

RCA supplies the CA3048 in a 16-pin dual in-line package.

An Audio Mixer

Where do you use four identical amplifiers? Figure 2-5 suggests one application, a four-channel audio mixer such as might be used in a recording studio or broadcasting station. Each channel of the mixer exhibits a high impedance—a minimum of 75K when the appropriate potentiometer is at its maximum-gain position. Each amplifier provides a gain of 20 dB when resistors R5 through R8 are 820Ω units, as shown. The gain can be varied by changing the values of these resistors. Figure 2-6 is a plot of amplifier gain versus input resistance for

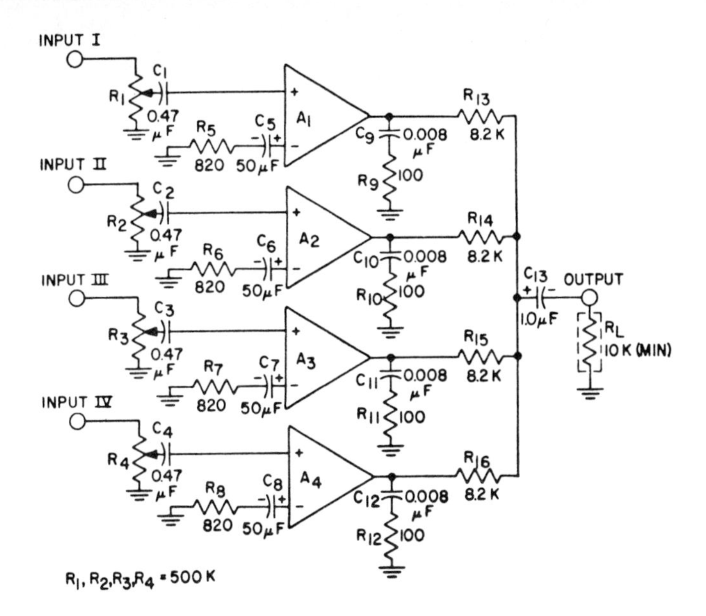

Fig. 2-5. A four-channel audio mixer using a single RCA CA3048. Each amplifier provides up to 20 dB gain with better than 45 dB isolation between channels.

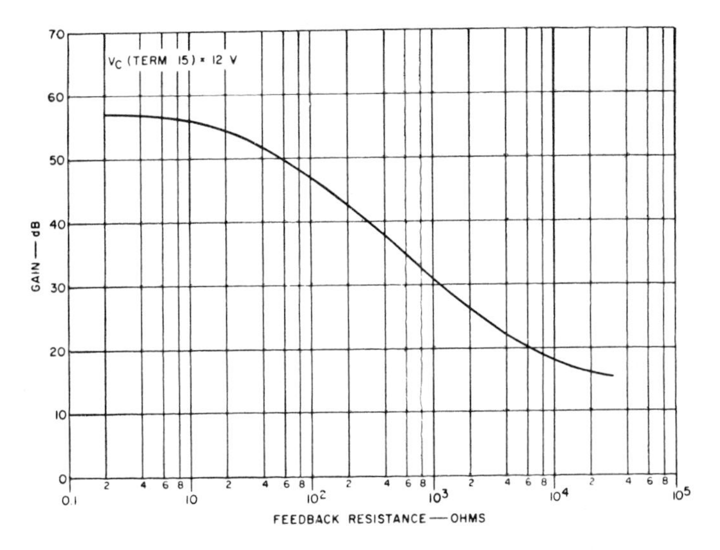

Fig. 2-6. Graph showing how the maximum gain of the amplifiers in Fig. 2-5 can be adjusted by varying resistors R5 through R8.

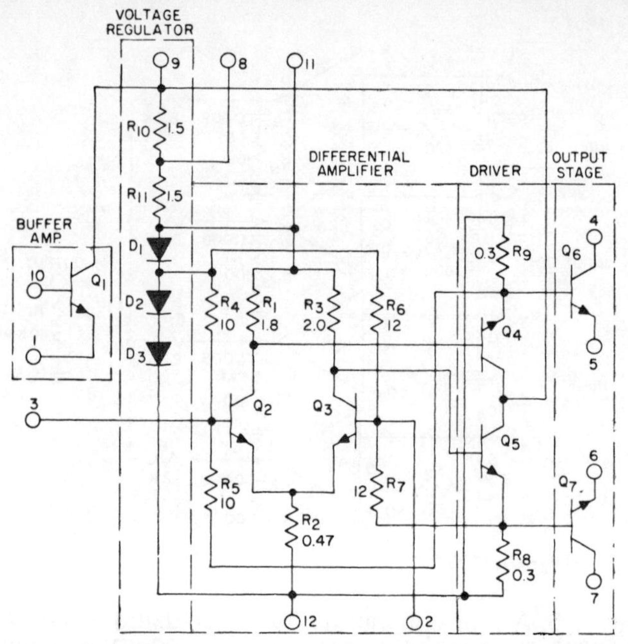

Fig. 2-7. The RCA CA3020 and CA3020A are designed to operate from single-voltage supplies. Both are supplied in 12-lead TO-5 cans. The CA3020 will deliver a maximum of ½W with a supply voltage of 9V maximum. The maximum supply voltage for the CA3020A is 12V. Its maximum power output is 1W.

a typical amplifier. To arrive at the gain for the mixer, the approximately 14 dB loss in resistors R13 through R16 is subtracted from the value found on the graph. The series resistor–capacitor combination on each output lead stabilizes the amplifier when then are operating into very high-impedance loads with minimum-gain control settings.

THE CA3020

The RCA CA3020/CA3020A power amplifier (Fig. 2-7) is a simple, inexpensive device, capable of delivering moderate power levels when fed by a3–12V supply (for the CA3020A only: : MAXIMUM SUPPLY VOLTAGE FOR THE CA3020A is 9V). Both devices exhibit gains of 75 dB or better, up to frequencies of 8 MHz. The only drawback to these amplifiers is their requirement for a split load. This comes ablut as a result of their push-pull output. The amplifiers are sold in 12-lead TO-5 cans.

As Fig. 2-7 shows, the basic circuit comprises a difference amplifier input stage whose double-ended output drives a

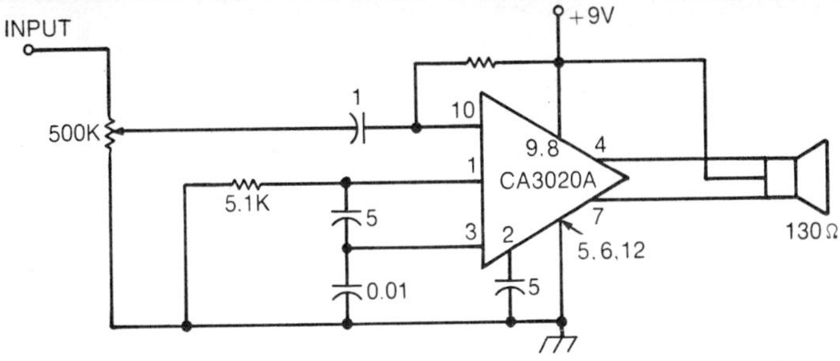

Fig. 2-8. Either a CA3020 or a CA3020A can be used to drive a speaker directly with up to 310 mW of power, if a speaker with a suitably high impedance is used.

matched pair of emitter followers. The output of the emitter followers is applied to the bases of two output transistors. The emitters and collectors of the output transistors are not connected to the rest of the circuit. The collectors must be connected to the V_{cc} supply through a suitable load, typically a center-tapped audio transformer. The emitters must be returned to ground, either directly or through small current-limiting resistors. An interesting feature of these amplifiers is the buffer stage, consisting of a single transistor, Q1. Generally, an input signal is applied between pin *10* and ground, and pin *1* is connected to pin *3* by means of a low-reactance capacitor. Thus, Q1 is used as an emitter follower to match a high impedance source to the relatively low impedance of this particular difference amplifier input.

Driving a Speaker Directly with the CA3020

A basic audio output stage using the CA3020 is shown in Fig. 2-8. Buffer transistor Q1 is biased by the 510K and 5.1K resistors connected to pins *10* and *1*. The signal is fed into pin *10* and coupled from pin *1* to pin *3* by means of a 3 μF capacitor. Pins *5* and *6*, connected to the emitters of the output transistors, are connected directly to ground; and pins *4* and *7*, the collector connections of the output transistors, connect to the outer taps of a rather unusual speaker voice coil. The speaker is RCA's 11113, with a 130Ω center-tapped voice coil—probably not something you have in your junkbox. Of course, a center-tapped transformer driving a speaker of more common impedance would work as well. Pins *8* and *9* are shorted together to bypass the 1.5K resistor in the bias voltage regulator. In this application, this increases the idling current in the circuit, moving the operating point of the output stage more into the class AB region and minimizing crossover distortion.

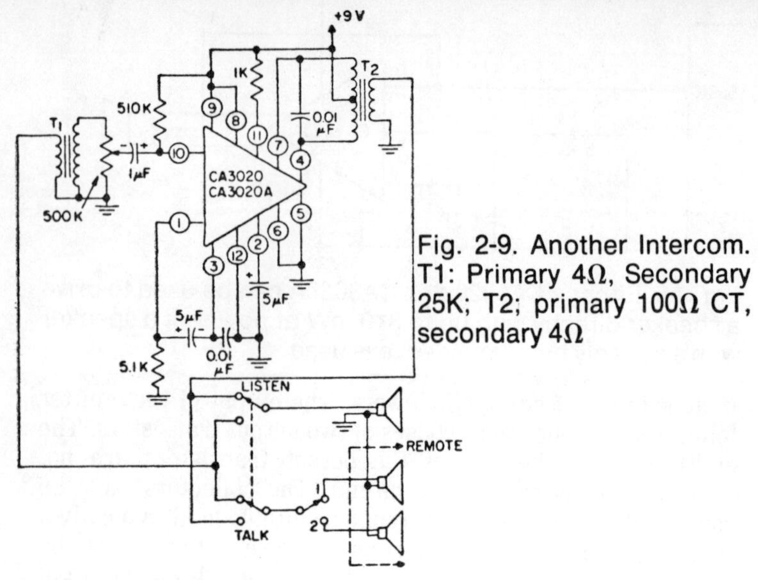

Fig. 2-9. Another Intercom. T1: Primary 4Ω, Secondary 25K; T2; primary 100Ω CT, secondary 4Ω

A CA3020 Intercom

The intercom in Fig. 2-9 features basically the same circuit as the one discussed earlier, with the trivial addition of input and output transformers to match 4Ω speakers. The only significant difference is the 1K resistor connecting pin *11* of the amplifier to the positive voltage supply. For the CA3020, this increases the idling current of the amplifier even more than merely shorting pins *8* and *9*, resulting in still less crossover distortion (and, of course, less gain).

THE LM380

National Semiconductor gets a lot of mileage out of a few transistors in its LM380 (Fig. 2-10). The device, supplied in a 14-pin DIP, is capable of a maximum power dissipation of 5W when heatsink pins *3*, *4*, *5*, *10*, *11*, and *12* (the middle three pins on either side of the package) are connected to a suitably large heatsink. With a pair of copper wings (such as the one in Fig. 2-11) soldered to these pins, the LM380 can dissipate up to 3.5W. An unusual input stage allows single-ended signals to be applied to either the inverting or the noninverting input. When this is done, the other input can be (1) grounded through a capacitor if the signal source impedance is high, or (2) grounded through a resistor equal to the signal source impedance if the input signal source has a low impedance. A fourth alternative, that of simply leaving the unused input floating, is satisfactory in many cases, as long as stray coupling doesn't cause the amplifier to go into oscillation.

That "unusual" input stage of the LM380 consists of a difference amplifier, each transistor of which is driven by an emitter follower. Comparing the LM380 with the RCA CA3020, which has an emitter follower driving one input device, we could say that National Semiconductor goes RCA one better in the input department. Of course, in defense of RCA, we can note that by making all three parts of their input transistor externally available, they offer the potential for more versatility.

The single-ended output from the LM380 difference amplifier input is applied to a common-emitter voltage gain stage and the output of this drives two transistors in what National Semiconductor calls a *quasi complementary-pair emitter follower*. By means of some interesting feedback, the gain of the LM380 is held to exactly 50 (34 dB). The feedback circuit includes the 25K resistor coming back from the output and the 1K resistor between the emitters of the difference amplifier transistors. The output of the amplifier varies around a null equal to half the supply voltage, thanks to the interaction of the 25K feedback resistor and the two 25K resistors in the voltage regulator section.

One More Intercom

Our first project illustrating use of the LM380 is yet another intercom (Fig. 2-12). Intercoms make good

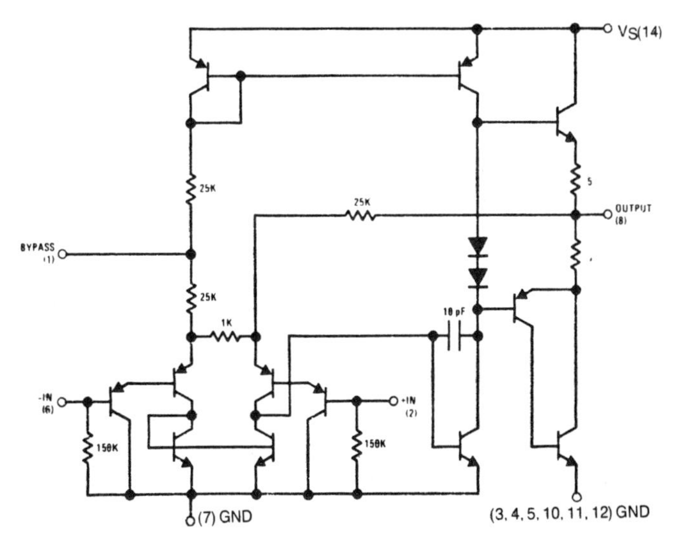

Fig. 2-10. The National Semiconductor LM380 is short on circuit complexity, long on power capability. With heat radiating fins (Fig. 2-11) it can deliver up to 3.5W of audio.

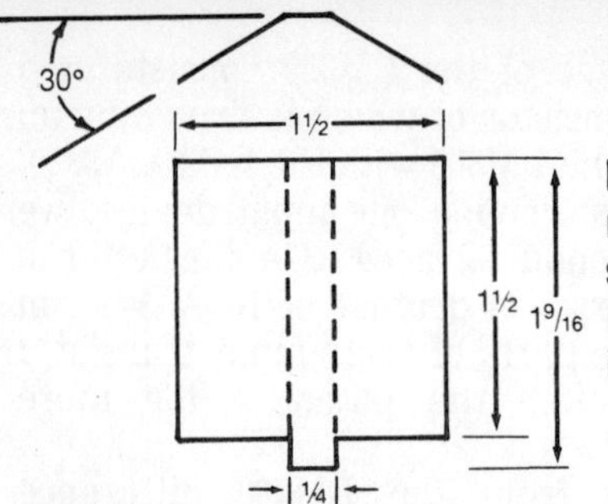

Fig. 2-11. With copper radiators like these soldered to the middle three leads on its DIP, the LM380 is rated up to 3.5W.

illustrations because they're simple yet presumably practical applications for audio power amplifiers. This intercom has an interesting feature in its "common-mode" volume control. With the wiper at the top of the pot, the inverting input is essentially operating at extremely low gain, while the gain at the noninverting input is maximum. As the wiper slides down the pot, the inverting mode gain increases, diminishing the overall gain of the circuit. Another way of looking at it is to say that as the wiper slides farther and farther down the pot, the input signal becomes more and more a common-mode signal.

All of the LM380 projects in this book show a phantom 2.7Ω resistor and 0.1 μF capacitor connected to the output of the amplifiers. These are needed to suppress a small oscillation in the 5 to 10 MHz range that sometimes occurs when the amplifier drives a low impedance load. You wouldn't, of course, notice such a high frequency oscillation in the speaker, but you might wonder what those new whistles were doing in your shortwave receiver.

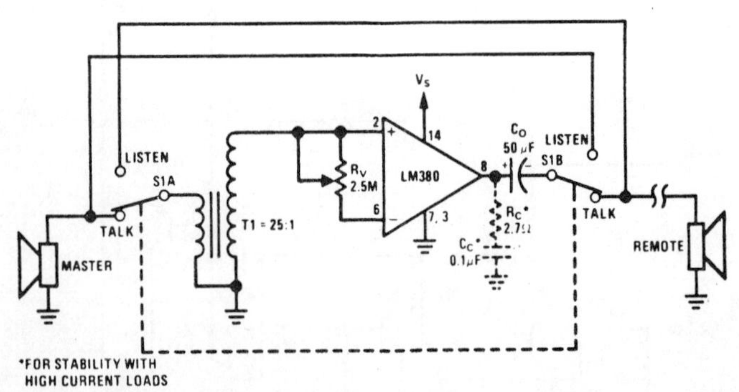

Fig. 2-12. Yet another intercom. This one features only one transformer. As explained in the text, R_c and C_c prevent RF oscillation of the amplifier; they are not necessary except when interference might result from oscillations. V_s equals 18V.

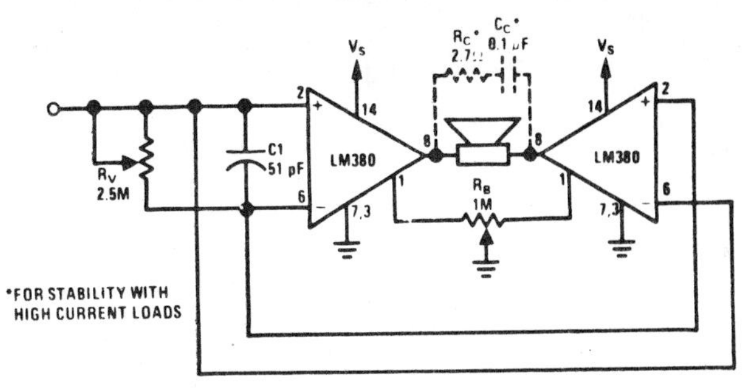

Fig. 2-13. Boosting power output by connecting two LM380 amps in a bridge arrangement. Resistor R8 balances current between the two amps when no signal is present. V_s equals 18V.

Bridge Amplifier

A few pages back, we used a pair of LH0021 amps in a bridge arrangement to produce higher gain than was available with a single amp. We can do the same thing with LM380s, as Fig. 2-13 shows. This is a simpler circuit than the one using the LH0021s. Using the "common-mode" volume control we discussed in the previous project, there is no necessity for having different gains for the two amplifiers. That's all taken care of by the volume control. The 1M resistor (R8) connected between pin *1* (bypass) terminals on the two amplifiers is necessary to zero the quiescent voltage drop between the two output terminals. If R8 were not there, there could be as much as 2V DC potential difference across the speaker.

RIAA Phono Amp

A step more complicated than a mere flat-response audio amplifier is the phono amplifier in Fig. 2-14. Phonograph records are recorded with low frequencies suppressed. This is necessary because of the thickness limitations of record groove walls. The function of attenuation vs frequency is the RIAA equalization curve. In order to play back a record and obtain a reasonable facsimile of the sound that was recorded on it, a phono amplifier must emphasize bass tones according to the RIAA characteristic. The additional components—that is, R1, C, and C2—approximate RIAA equalization for the amplifier in the figure. In terms of a Bode frequency plot, gain is flat up to about 60 Hz, where it starts to drop at 6 dB per octave to about 500 Hz. Then it flattens out until it reaches about 2 kHz, where it again resumes its drop of 6 dB per octave.

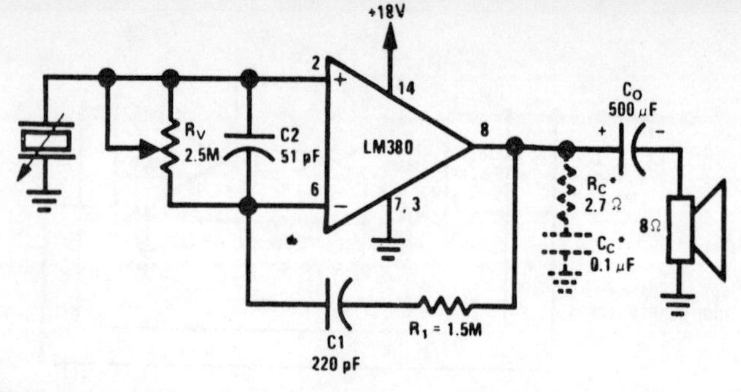

Fig. 2-14. With only a handful of components, a the LM280 becomes a complete phono amplifier with RIAA equalization.

THE LM381

National Semiconductor's LM381 (Fig. 2-15) was designed as an extremely low noise (1 to 1.6 dB noise figure, depending on load resistance), high gain (112 dB) dual preamplifier. Besides its low noise figure, the LM381 features 60 dB isolation between channels and 120 dB rejection of signals coupled through the power supply. It is designed to operate with single-voltage supplies of 9 to 40V. National offers this little goody in a 14-pin DIP.

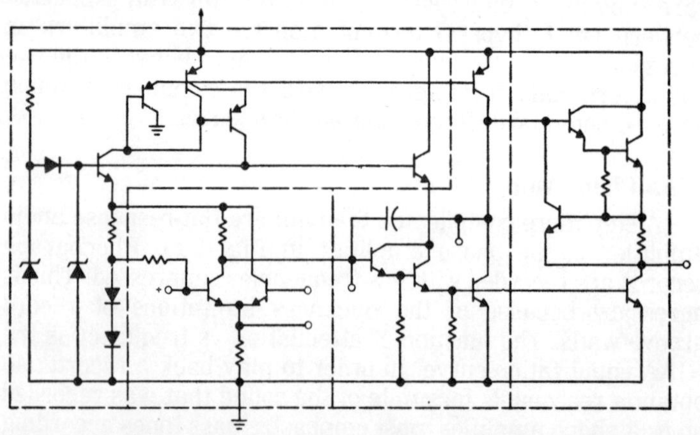

Fig. 2-15. The National Semiconductor LM381 is a dual preamplifier with an extremely low noise figure. The circuit diagram above represents one of the two amplifiers on the IC. Where terminals are identified by a pair of numbers in parentheses, the first number in each set refers to one amplifier and the second number refers to the other. A dual amp such as this is ideal for stereo applications.

The circuit consists of a difference amplifier that can be operated single-ended for minimum noise figure. For single-ended operation, the signal is fed to pin *1* or *14* and pin 2 or *13* is connected to ground. In this case, transistor Q1 operates as an ordinary class A common-emitter gain stage. Transistor Q5 provides another stage of common-emitter gain. It is driven by a darlington emitter-follower pair. The output stage is another darlington emitter-follower pair consisting of Q8 and Q9. Transistor Q10 limits short-circuit current to 12 mA.

NAB Tape Playback Amp

The combination of low noise figure and high gain makes the LM381 ideal for use as a tape playback preamplifier. As was the case with the preceding phono amplifier, tape amps also require equalization, although for slightly different reasons. The circuit in Fig. 2-16 provides a response that closely matches the NAB equalization characteristic. There is little remarkable about the arrangement. The feedback network, consisting of the 2.2M resistor, the 62K resistor, and the 1500 pF capacitor, works with the source network connected between pin 2 or *13* and ground to provide the necessary shaping of the gain characteristic. The playback transducer is not precisely specified, but this circuit should work well with a variety of tape heads.

Playback Amp with Fast Turn-On

The one disadvantage of the circuit in Fig. 2-16 is that it takes about 5 seconds from the time that power is applied to the time the circuit is operational. This is the time it takes for

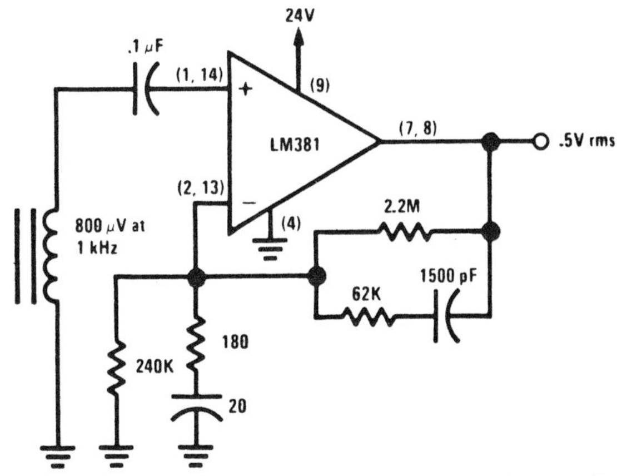

Fig. 2-16. A tape playback amplifier with NAB equalization.

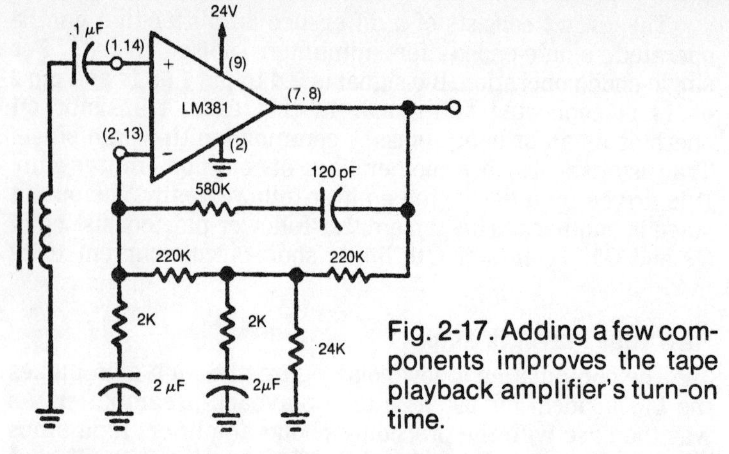

Fig. 2-17. Adding a few components improves the tape playback amplifier's turn-on time.

the 20 μF capacitor to charge through the 2.2M resistor. The turn-on delay of the circuit in Fig. 2-17, however, is only 0.1 second. This improvement comes about through the use of a smaller RC time constant. There are really only two design differences between this circuit and the previous one. The first is simply a reduction by a factor of ten in the size of the resistors controlling the gain of the amplifier. The second is in the use of a more complex filter for controlling the low-frequency response. This increase in complexity allows a decrease in capacitor size.

NAB Tape Record Amplifier

If the LM381 is suitable for use as a tape playback amplifier, it should be also suitable for use as a record amplifier. Figure 2-18 shows just such a circuit. The basic noninverting-mode op-amp design hasn't changed much, although the pole frequencies of the input filter are now wide apart and the feedback element is a pure resistance. As the caption notes, L1 and C6 are tuned to keep the bias signal out of

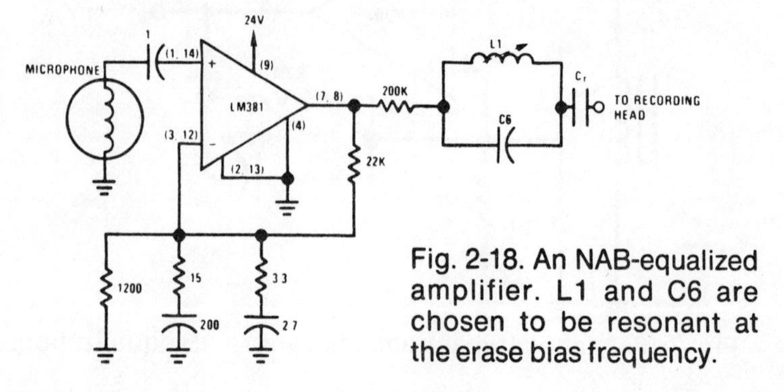

Fig. 2-18. An NAB-equalized amplifier. L1 and C6 are chosen to be resonant at the erase bias frequency.

the amplifier, where it could lead to intermodulation distortion. (Tape recorders apply a barely supersonic "bias" signal to the record heads along with the signal that is actually to be recorded. The bias keeps the oxide particles on the tape from arranging themselves into noisy patterns during quiet periods.)

Magnetic-Pickup Phono Preamp with Tone Control

Figure 2-19 illustrates a final, elegant use of the LM381, a complete magnetic phono preamp with tone control. The gain control circuit resembles that for the slow turn-on tape playback amplifier, except for a more elaborate feedback network. The tone control network to the right of the 1 μF capacitor is a classic tone control design.

THE LM170

So far, the IC amplifiers we have looked at have had their gain set at fixed levels by means of external feedback components. National Semiconductor's LM170 (Fig. 2-20) is a high-gain IC amplifier with built-in AGC and squelch. That is, it provides an automatic gain control to give optimum output regardless of input level and squelch to run the gain to zero in the absence of an input signal. The LM170 is designed to operate from a single-voltage supply of from 6 to 24V. Nominal supply voltage is 12V. National Semiconductor offers the amplifier in both a 14-pin DIP and a 10-pin TO-5 package.

The input stage in the circuit of Fig. 2-20 is a basic difference amplifier with double-ended output. The difference amplifier consists of transistors Q1 and Q2 which feed lateral *pnp*'s Q12 and Q13. The next stage is the gain control section. The double-ended signal from the input stage is fed to the bases of Q3 and Q6. An AGC feedback signal is applied to the bases of Q4 and Q5. Transistors Q3 and Q4, together with Q5

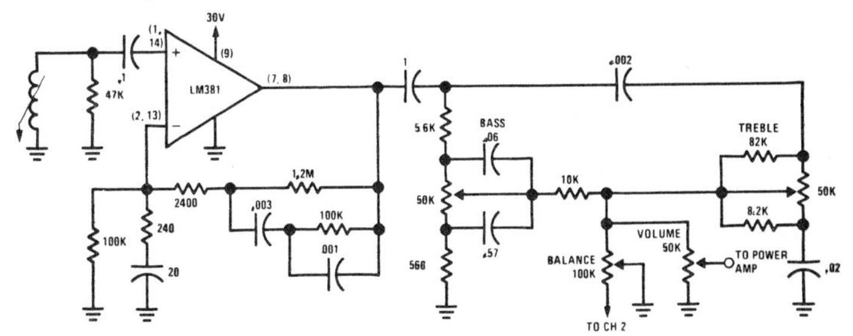

Fig. 2-19. Going all the way. One channel of a complete stereo preamp. The other channel uses the other amplifier of the LM381.

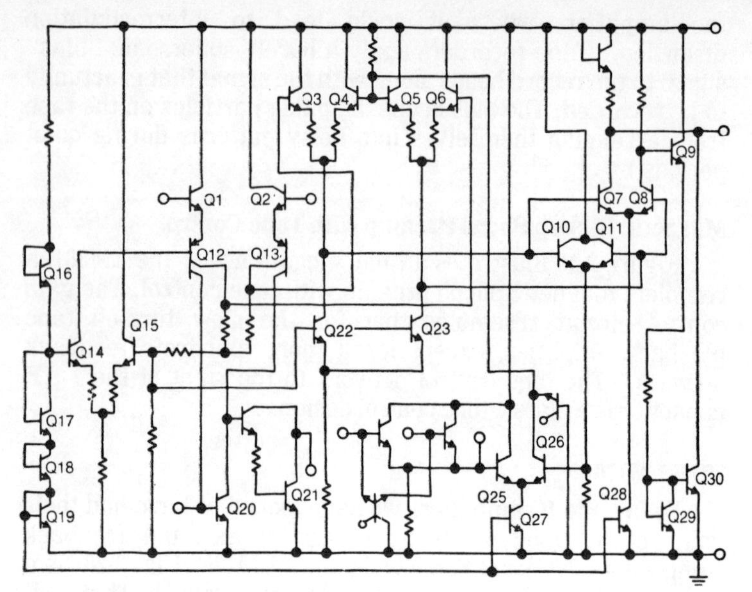

Fig. 2-20. The National Semiconductor LM170 AGC/squelch amplifier looks complex and it is. Stage-by-stage explanation is offered in the text.

and Q6, form differential pairs sharing common current sources. The current source for Q3 and Q4 is Q22, while Q23 is the current source for Q5 and Q6. How does all this work to provide AGC? When there is no AGC signal, Q4 and Q5 are cut off and Q3 and Q6 form a basic difference amplifier with double-ended output. When an AGC signal is applied to Q4 and Q5, these transistors start to conduct, robbing current from the differential amplifier transistors and reducing gain. Where does the AGC signal come from? There are two choices. Staying entirely within the IC, Q10 and Q11 provide a level shift for the signal, which is then amplified by Q26 and fed back to Q4 and Q5. If more AGC is required, a signal from the output of the LM170 can be fed back externally to pins 2, 3, or 4.

The output stage consists of difference amplifier Q7 and Q8 driving common-emitter voltage amplifier Q30 through emitter follower Q9.

What about squelch? Going back to the input stage, we notice that the second collectors of lateral pair Q12 and Q13 drive the amplifier consisting of Q20, Q36, and Q21. The collector of the darlington pair output stage is available at pin 6, identified as SQUELCH OUT in the figure. When squelch is desired, pin 6 is connected to the positive supply voltage through a load resistor, to ground through a capacitor, and to

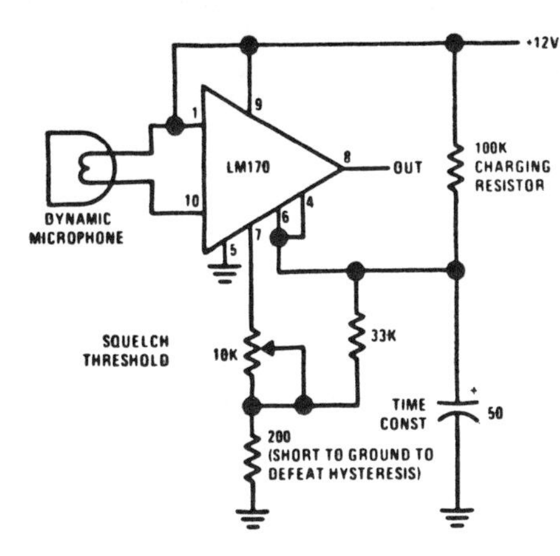

Fig. 2-21. Fast-attack, slow-decay characterize the squelch action of this circuit. There's no "pop" as this squelch cuts off.

the AGC stage via pin 4. The RC time constant of the two external elements controls the slow-decay characteristics of the squelch. Squelch threshold is adjusted by varying the bias on the base of Q20 by means of an external pot connected between pin 7 and ground.

Preamp with AGC and Squelch

For a simple application of the LM170, let's look at Fig. 2-21. The microphone is a single-ended input, so pin *1*, the noninverting input, is grounded through the power supply. The squelch time constant is set by the 100K resistor and the 50 μF electrolytic capacitor. The squelch in this project has some built-in hysteresis, caused by feeding back some of the squelch output to the squelch threshold.

What applications does such an amplifier have? Some places that would find use for several would be courtrooms or city council meeting chambers where several microphones feed a common power amplifier. If each microphone were equipped with an amplifier like the one in Fig. 2-21, it would be insensitive to background noise, responding only to sounds spoken directly into it. Moreover, the output level would be maintained at a reasonable volume, regardless of whether the speaker whispered or shouted. Another application in which these characteristics would be useful would be in an aircraft intercom.

Voice-Operated Switch

There are applications, such as voice-actuated tape recorders, and communications transceivers, in which a voice-operated relay (VOX) is desirable. The LM170 squelch output can be used to drive such a relay through a suitable

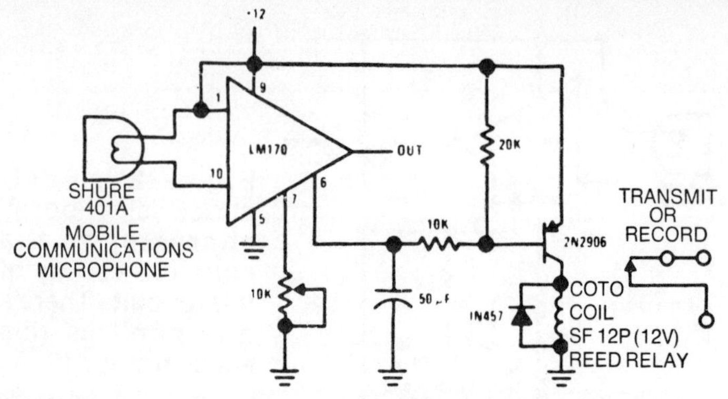

Fig. 2-22. For VOX (voice-operated relay) applications, the LM170's squelch output can drive a transistor switch.

transistor, as in Fig. 2-22. The circuit needs little explanation, since it is much the same as the preceding example (minus the squelch function and the hysteresis). The diode suppresses the noise spikes that occur when the transistor shuts off and the coil tries to maintain its current flow.

THE CA3094

Another means of controlling amplifier gain by means of an external signal is provided by the RCA CA3094 (Fig. 2-23). By changing the bias at the base of the constant-current source (transistor Q3), the total current shared by Q1 and Q2 can be varied, thus varying the gain. The double-ended output of this stage is applied to a darlington-pair difference amplifier, the single-ended output of which drives the final darlington-pair power amplifier. The collector and emitter of the output transistor are both externally available, allowing the device to either *source* or *sink* the external load. The input to the final stage is available via pin *1* for either frequency compensation or for shutting off (inhibiting) that stage.

The CA3094 can use either single- or dual-voltage supplies. Typical dual supply potential is ±12V and single supply potential is typically 24V. The amplifier can provide 0.6W of audio output when operating class A. Gain is typically 100 dB. An eight lead TO-5 package is standard.

Phono Amplifier with Feedback Tone Control

The complete phono amplifier in Fig. 2-24 is like the preamplifier in Fig. 2-19, but with an important difference. The tone control in the earlier example merely attenuated selected ranges of frequencies in RC networks. Tone control in this circuit is achieved by selective variation of the feedback

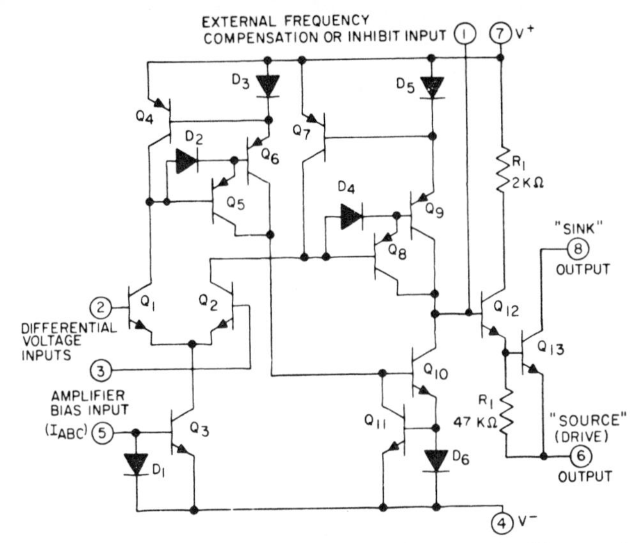

Fig. 2-23. The RCA CA3094 allows for controlling gain by controlling the bias current into pin 5. It can be operated with either a dual- or a single-voltage supply. It is supplied in an eight-lead TO-5 can.

impedances. Note that the bass and treble networks run from the output of the transistor output stage back to the inverting input. We look more deeply into this kind of frequency response shaping in the chapter on filters.

The IC portion of the circuit in the figure. is straightforward. Feedback is applied to the inverting input and a single-ended input signal is applied to the noninverting input. The bypass capacitors for the power supply leads are both shown. A small capacitor from the output back to terminal *1* provides compensation. The open-loop gain is fixed by the bias resistor on pin *5*.

The output darlington pair is connected as a common-emitter voltage amplifier. A 1Ω resistor in the emitter lead limits no-load current.

The amplifier output stage uses a matched pair of complementary-symmetry discrete power transistors connected as emitter followers to drive R_L, an 8Ω speaker. As noted earlier, part of the output of this stage is fed back for tone control.

The dual-voltage power supply in the figure is deserving of a word or two in passing. It might appear at first that the electrolytic capacitors on the power supply output are a bit of an overkill. After all, 9400 μF is a lot of capacitance. The

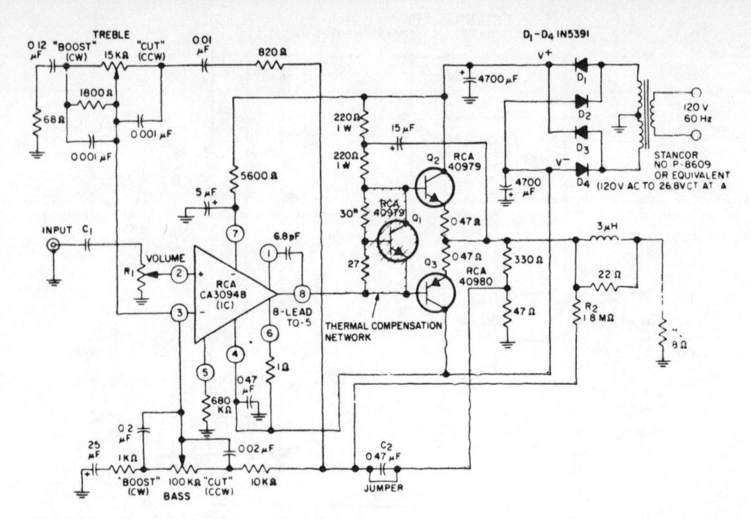

Fig. 2-24. A 12W hi-fi amplifier, with tone controls using a CA094 to drive a matched complementary pair of power transistors. Note simple power supply filter requirements. Ceramic cartridge input: C1, 0.0047 μF; R1, 250K (short C2 and R2).

reason these capacitors are as big as they are, however, is because they have another function besides power supply filtering. During low- and moderate-level recording passages, the capacitors store up energy. During very loud passages, they give up this energy to the final amplifier. By storing energy in the capacitors this way, we are able to get by with a smaller power transformer than would otherwise be necessary. The result is a lighter, more efficient power supply.

Oscillators and Signal Generators

The oscillators and generators in this chapter for the most part have square-wave outputs. From an RF standpoint, there are times when a square-wave output is desirable and times when it isn't. In an application like a frequency/marker generator, a square-wave output, rich in harmonics, is desired. Where a voltage-controlled oscillator (VCO) is to be used in a phase-locked loop, a square wave is also desired. However, where the output of a generator is to be used to drive a transmitter, harmonics are undesirable. In these cases, the generator output can be filtered to remove all but the fundamental frequency.

THE LM139 COMPARATORS

Many of the projects that follow involve comparators. The basic comparator circuit consists of an op-amp with positive feedback and a reference voltage applied to the noninverting input. When the voltage at the inverting input is less than the reference voltage, the output of the comparator is high. When the voltage at the inverting input exceeds the reference voltage, even by a little bit, the output of the comparator switches immediately to low. The key to this very fast shift of state is the positive feedback. Positive feedback assures that the gain around the amplifier loop will be so high that all transistors in the amplifier will be fully shut off or fully saturated.

Linear ICs intended for duty solely as comparators can be of simpler design than op-amp ICs, since they do not require compensation or excessive regulation. One such linear IC is National Semiconductor's LM139 comparator. The LM139, with four comparators on a single IC chip, is designed to operate from either single- or dual-voltage supplies. Supply voltages can range from ±1V (+2V on a single-voltage supply) to ±18V (+36V on a single-voltage supply). The LM139 is available in 14-pin DIPs and flat-packs. Where a project does not use all four of the LM139's comparators, the manufacturer warns us that all input and output pins of unused comparators should be grounded.

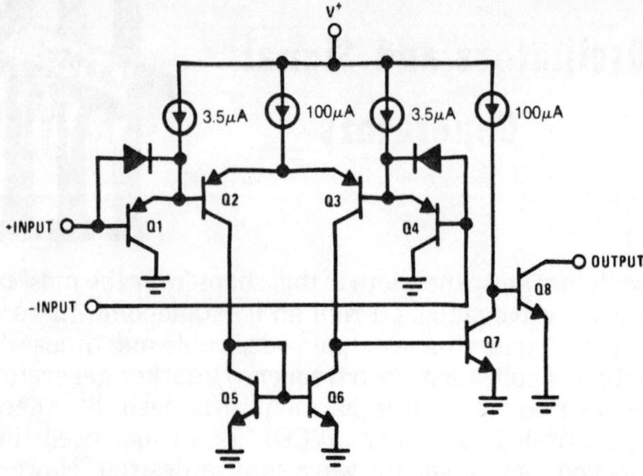

Fig. 3-1. A simplified circuit diagram of one of the four comparators on the LM139. The output arrangement is typical of comparators. Q5 can be used to switch some external circuit or it can be connected through a pullup resistor to the positive voltage supply in order to provide a logic output.

A simplified circuit for one LM139 comparator is shown in Fig. 3-1. The circuit is simply a darlington-pair difference amplifier with single-ended output driving a common-emitter darlington output stage. The collector of the output transistor is connected to a pin on the package. The LM139 can be used either to switch an external load by means of this output transistor or, if the collector of the transistor is connected to a positive voltage through a *pullup* resistor, it can provide a logic output. That is, when the output transistor is cut off, the output (seen by a high-impedance load) will be the voltage applied to the pullup resistor. When the output transistor is saturated, the output will be at ground potential. The pullup resistor limits the current flow through the transistor when it is conducting.

Basic Square-Wave Generator

For a practical illustration of a comparator used as an oscillator, consider Fig. 3-2. When power is initially applied to the circuit, the output transistor is cut off, the potential at V_o is V+, and the potential at the noninverting input is two-thirds the value of V+. Immediately, the 75 pF capacitor begins charging through the 100K resistor. When its voltage exceeds the voltage on the noninverting terminal, the output transistor saturates, pulling V_o to zero. This changes the reference

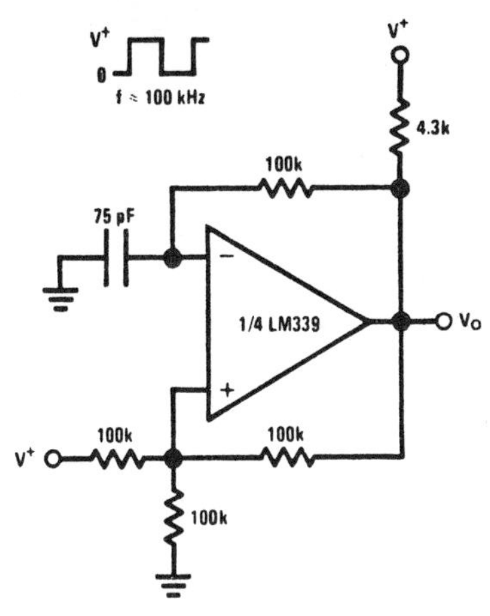

Fig. 3-2. Basic square-wave generator with single-voltage supply. When the output transistor of the LM139 is nonconducting, the 4.3K resistor "pulls up" the output voltage to V+. When the output transistor conducts, the output falls to ground. The 100K positive feedback resistor provides hysteresis so that reference voltage oscillates between one-third and two-thirds of V+.

voltage, lowering it to one-third the value of V+ and assuring that the amplifier is in full saturation. Now the capacitor *discharges* through the resistor until its potential falls below the new reference voltage, at which time the output changes state again. In this way, the voltage across the capacitor oscillates in a sawtooth fashion between one-third and two-thirds of the applied voltage, and the comparator output is a square wave that oscillates between the applied voltage and ground. Oscillating frequency can be calculated from

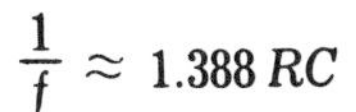

$$\frac{1}{f} \approx 1.388\,RC$$

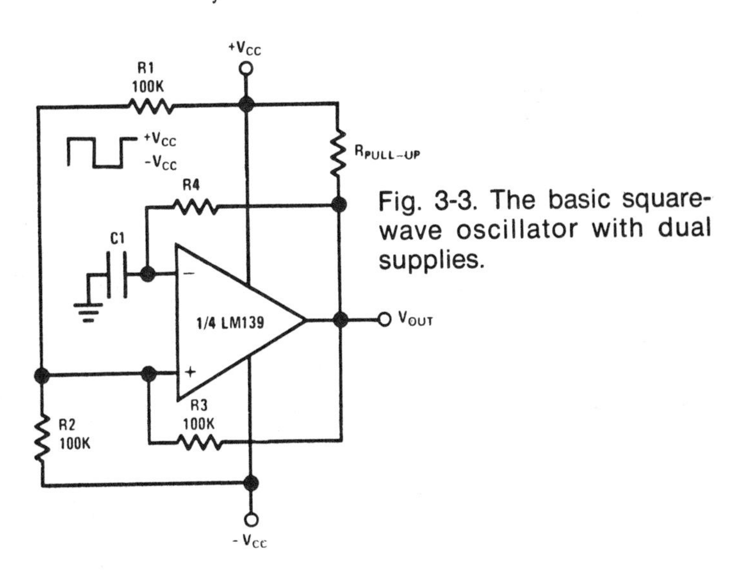

Fig. 3-3. The basic square-wave oscillator with dual supplies.

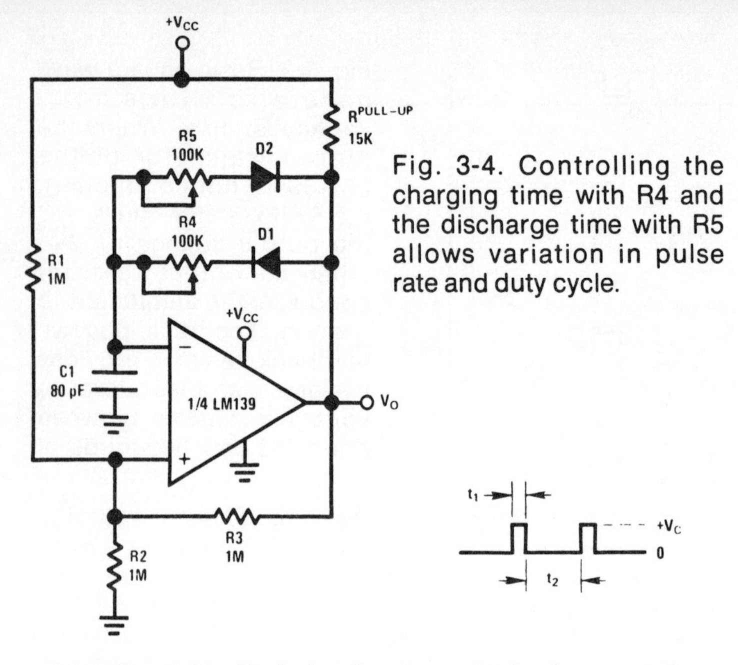

Fig. 3-4. Controlling the charging time with R4 and the discharge time with R5 allows variation in pulse rate and duty cycle.

An almost identical circuit using a dual-voltage supply is shown in Fig. 3-3. As the figure suggests, its output oscillates between $\pm V_{cc}$.

Variable Frequency and Duty Cycle Generator

The preceding two square-wave oscillators produce a string of pulses with a 50% duty cycle; i.e., time on is equal to time off. The ratio of pulse width to pulse spacing can be varied with a modified circuit like the one in Fig. 3-4. Opposing diodes direct the charging and discharging currents of the capacitor through different potentiometers to change the RC time constant during charge and discharge. Potentiometer R4, controlling charge time, affects pulse duration tt_1.1p Potentiometer R5, controlling discharge time, affects the spacing t_2 between pulses.

The feedback resistors in this circuit have been increased by a factor of ten over those in the earlier circuits because the size of the pullup resistor has been increased. The potential across the capacitor still oscillates between 0.3334 and 0.6667 V_{cc}, since R4 and R5 are identical.

Crystal-Controlled Square-Wave Generator

Oscillator frequency can be controlled with a crystal, as in Fig. 3-5. Note that the positions of the inverting and noninverting inputs on the drawing have been reversed relative to the preceding drawings. The circuit, however, remains the same, except that the feedback resistor has been replaced by a crystal. The oscillating frequency of R3 and C1 is selected to be several times higher than the crystal frequency. In operation, this RC combination acts as a filter, maintaining a DC voltage at the inverting terminal equal to half the supply voltage. this is the average DC value of the output signal. The crystal acts as a frequency-selective feedback device, causing the voltage at the noninverting terminal to oscillate between V_{cc} and $0.5V_{cc}$ at a rate determined by the crystal characteristics. This should be a series-resonant type crystal. That is, it should present a short circuit to signals at its resonant frequency.

Voltage-Controlled Oscillator

A simple voltage-controlled oscillator can be had using three of the four comparators of the LM139 (Fig. 3-6). To understand its operation, assume that when power is applied, comparator 2 is driven to positive saturation and its output is V_{cc}. This shuts off the output transistor in comparator 3 and prevents any current from flowing through the 50K resistor to ground. The *control* voltage then begins to charge capacitor C1 at a linear rate, due to the integrating action of comparator 1, which is being used in its linear operating region. When the voltage across C2 reaches a minus value of 0.5 V_{cc} (15V), comparator 2 trips, bringing the output voltage to zero. The output of comparator 3 likewise goes to zero, turning its output

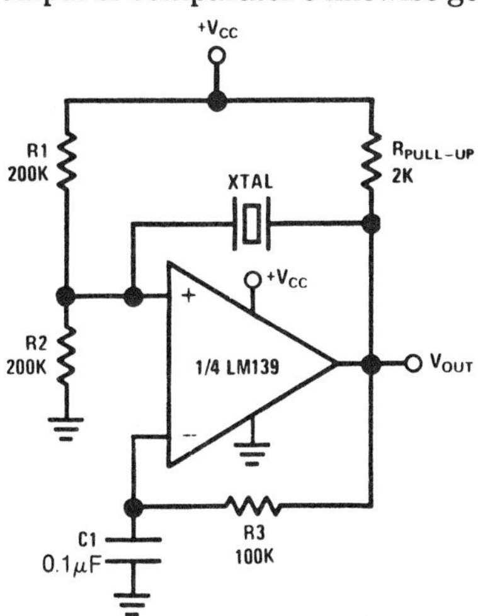

Fig. 3-5. A series-resonant crystal in the positive feedback loop produces oscillations. R3 and C1 form a filter that maintains the voltage at the inverting input at a level equal to the DC average of the output.

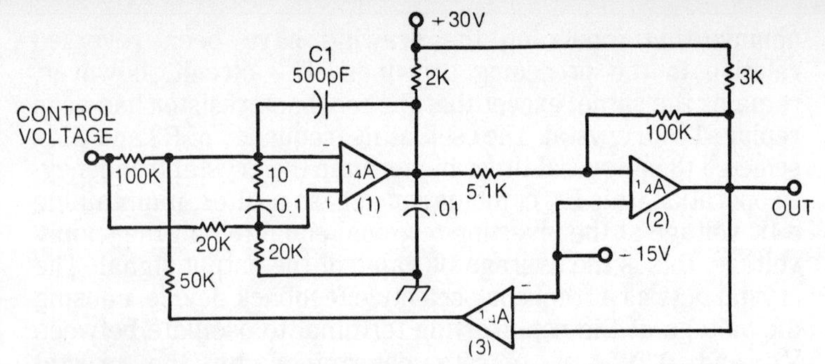

Fig. 3-6. A voltage-controlled oscillator. Operation is explained in the text.

transistor full on and allowing C1 to discharge through the 50K resistor. The output of comparator 1 now increases, and when it reaches +15V, comparator 2 switches back to a positive output, bringing the oscillation full cycle.

For the circuit values given, frequency range is from about 670 Hz to 115 kHz, the control voltage varying from 250 mV to 50V.

THE CA3000

The RCA CA3000, although intended as an operational-type amplifier, is another simple IC. Among its uses is voltage comparison. As Fig. 3-7 indicates, the CA3000 is a single darlington difference amplifier with double-ended output. The base of the constant-current generator transistor is externally available, and the gain of the amplifier can be varied by varying the bias into this port. Typical supply voltages are ±6V. The amplifier, which cannot really be called an op-amp because of its moderate 30 dB gain, is available in a 10-pin TO-5 can.

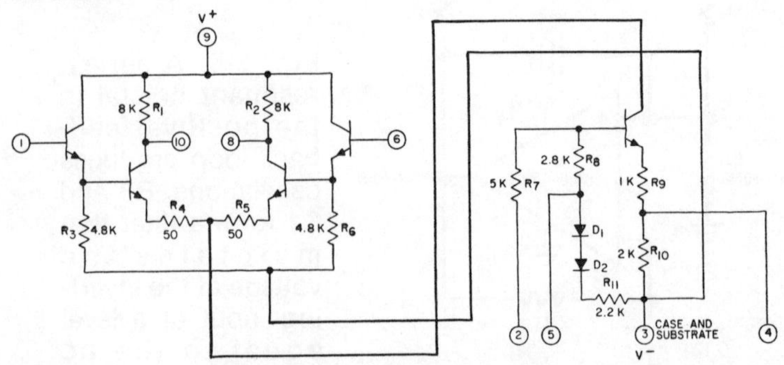

Fig. 3-7. RCA's CA3000, a difference amplifier whose gain can be controlled by varying the bias current into pin 2.

Modulated Oscillators

The crystal oscillator in Fig. 3-8 can be modulated by applying a signal to pin 2, the base connection for the current generator of the CA3000. Even though the gain of the CA3000 is low for an op-amp, it is still adequate to allow the amplifier to function as an ordinary feedback-type oscillator. Single-ended output is taken from one collector of the difference amplifier and fed back to its inverting input through a series-resonant crystal. Modulated sine-wave output is taken from the other difference amplifier collector. As the 100K resistor suggests, this circuit requires a high-impedance load.

Another modulated oscillator (Fig. 3-9) uses the National Semiconductor LM170 we encountered in Chapter 2. In this application, the squelch function of the LM170 is not used, but a modulating signal *is* applied to one of the AGC inputs. Here, the feedback element, a ceramic filter, is in the inverting loop, so it must be parallel resonant, feeding back at all frequencies *except* its resonant frequency. Signal level is controlled by varying a bias level controlled by the 100K potentiometer, which is applied along with the modulating signal. For stability, part of the output signal is fed back to the AGC input throught the 100 pF capacitor. This signal is peak detected in the AGC circuit, filtered by the 0.1 μF capacitor connected to pin 2, and applied to the gain-control amplifier along with the modulation and level inputs.

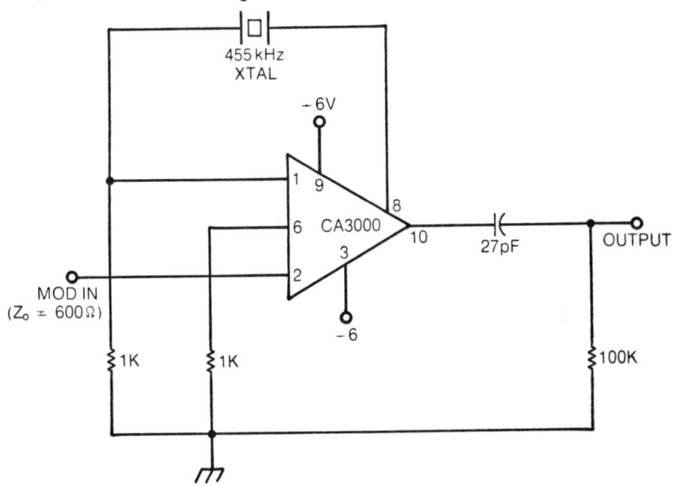

Fig. 3-8. The gain-control feature of the CA3000 lets us modulate this crystal oscillator. Such a circuit would be useful as an alignment generator for receivers with a 455 kHz intermediate frequency.

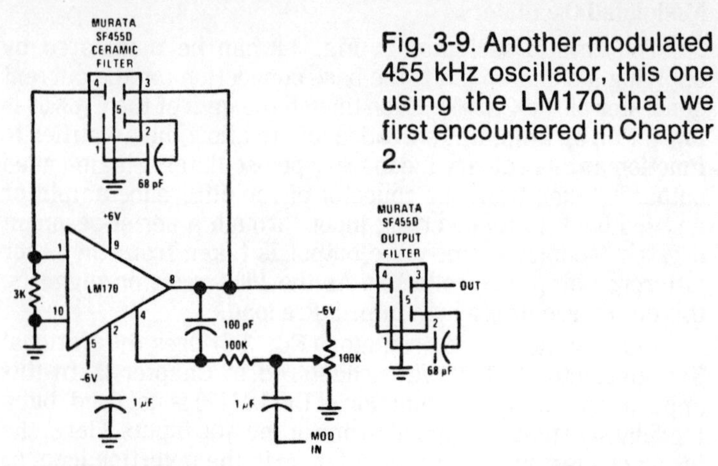

Fig. 3-9. Another modulated 455 kHz oscillator, this one using the LM170 that we first encountered in Chapter 2.

Twin-T Oscillator

A lower frequency oscillator along the same lines is shown in Fig. 3-10. The feedback element is a twin-T network, tuned by means of the 25M variable resistor. Like the ceramic filter in the previous project, the twin-T network feeds a signal back to the inverting input at all frequencies except its resonant frequency. Of course, for the component values given, that frequency is much lower than the 455 kHz of the ceramic filter. The output of this oscillator is tunable over more than a decade, from 320 Hz to 3300 Hz. To make sure that the circuit oscillates, wideband positive feedback is provided

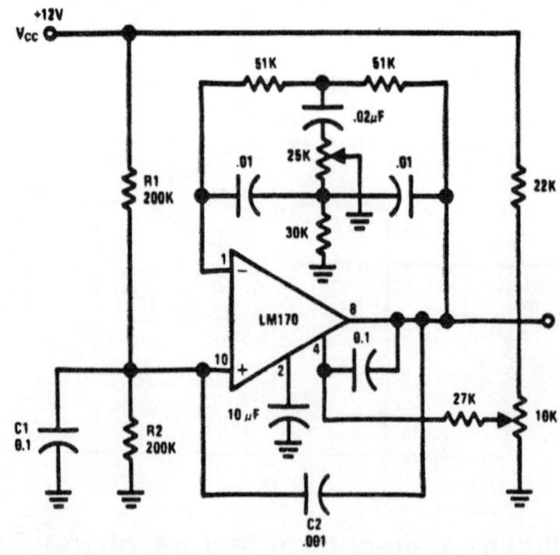

Fig. 3-10. This oscillator replaces the ceramic filter of the previous project with a twin-T filter to achieve lower output frequencies.

by feedback capacitor C2 and source capacitor C1. Some AGC is obtained by feeding back a portion of the output via the 0.001 μF capacitor to pin 4, and filtering the peak detected output with the 10 μF capacitor at pin 2. The potentiometer sets output level.

Wien Bridge Oscillator

For even lower frequencies, the Wien bridge oscillator is an old standby. Figure 3-11 shows a Wien bridge oscillator with an output frequency of either 8 or 80 Hz, depending on the values selected for C1 and C2. Gain of approximately four is assured by the feedback and source resistors connected to the inverting input, and the Wien bridge network is connected to the noninverting terminal. This network consists of the series resistor and capacitor, R1 and C1, and the parallel network consisting of R2A, R2B, and C2.

THE CA3033/CA3047

In contrast to the linear ICs described so far in this chapter, RCA's CA3033/CA3047 is a bonafide operational amplifier. Moreover, it is capable of high output currents—up to 36 mA with a dual-voltage ±12V supply. The chip in both ICs is identical; the only difference between the CA3033 and the CA3047 is the package. The former is packaged in a 14-pin ceramic DIP, while a plastic package is used for the latter.

Figure 3-12 shows the circuit for the CA3033/CA3047. The input is a dual emitter-follower driven difference amplifier. This is followed by a similar stage with single-ended output. The output stage is a single-ended class B power amplifier

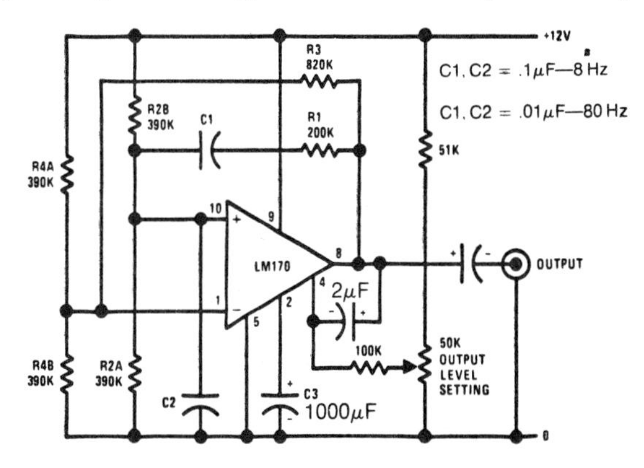

Fig. 3-11. Still lower output frequencies, down to 8 Hz, require a Wien bridge oscillator such as this.

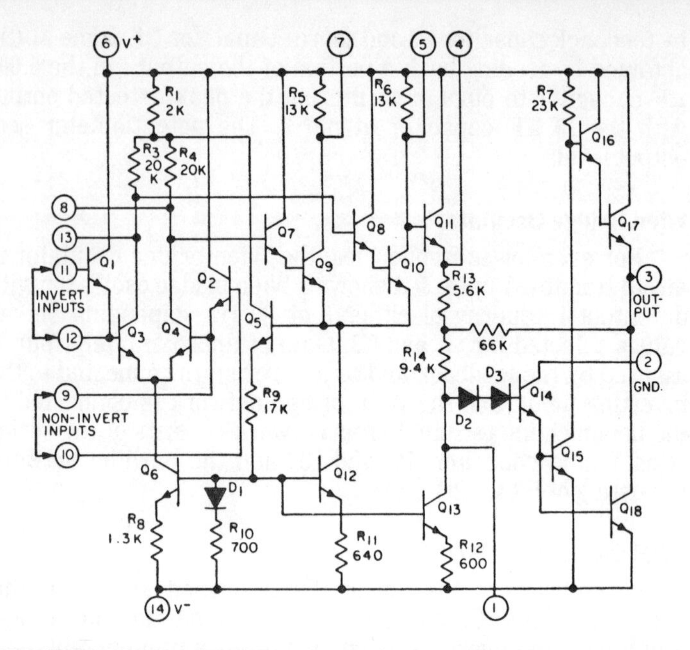

Fig. 3-12. RCA's CA3033 and CA3047 share this same chip, the only difference between the two being in their packages. Needless to say, rules for numbering ICs could stand some standardization.

consisting of Q17 and Q18. The output stage can be cut off by a positive voltage applied to pin *1*. Pins *7*, *5*, *8*, and *13* are available for compensation.

The square-wave generator in Fig. 3-13, using the CA3033, is essentially the same as the one in Fig. 3-3, except for the provision for a squelch. If R1 and R2 are the same size, say 1M each, the voltage on the capacitor will oscillate between 0.5 V_{cc} and 0.5 V_{ee}. The approximate oscillation frequency may be calculated using

$$\frac{1}{f} \approx 2.2\,RC$$

Voltage-Controlled Oscillator

We encountered RCA's CA3094 in Chapter 2. Figure 3-14 repeats the CA3094 input stage, which has some resemblance to the CA3000. The current shared by the difference pair is controlled by Q3, which can itself be controlled by an outside bias, applied via pin 5. Figure 3-15 shows how this outside bias can be used to obtain a voltage-controlled oscillator of a sort. The circuit is the basic comparator oscillator with which we

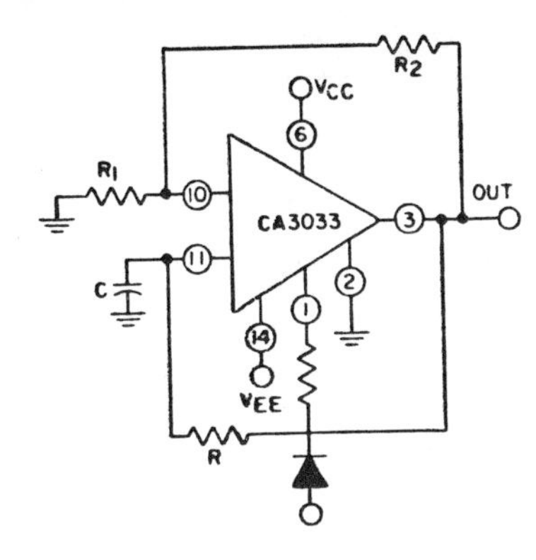

Fig. 3-13. This standard square-wave generator can be squelched by applying a positive voltage to pin 1.

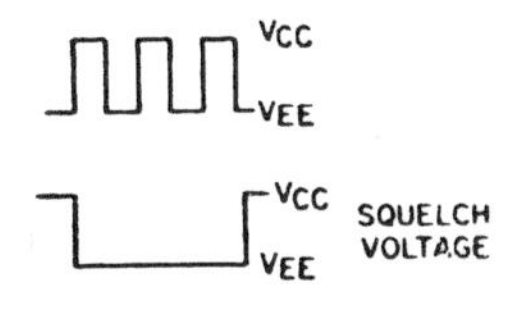

are by now familiar, with the addition of a diode in series with the resistor through which the capacitor charges. As in the variable oscillator of Fig. 3-4, this allows the capacitor to charge through the 20K resistor, but prevents it from discharging through it. How then does the capacitor discharge? The answer is through Q2 and Q3. But while Q2 is driven into saturation, Q3 is operating in its linear region, limiting the discharge current to a value determined by its bias level. The notes on the figure give the frequencies corresponding to minimum and maximum current and voltage bias inputs. If the available voltage for controlling the frequency were other than 13.5V, the value of the resistor could be changed, the only requirement being that it limit the current into pin 5 to a maximum of 500 μA.

The output of this oscillator does not have a 50% duty cycle, except at the frequency where the capacitor's charging and discharging currents are exactly equal. A flip-flop on the output of the oscillator would even up on the *on* and *off* times of the pulses, but would, of course, also divide the frequency by two.

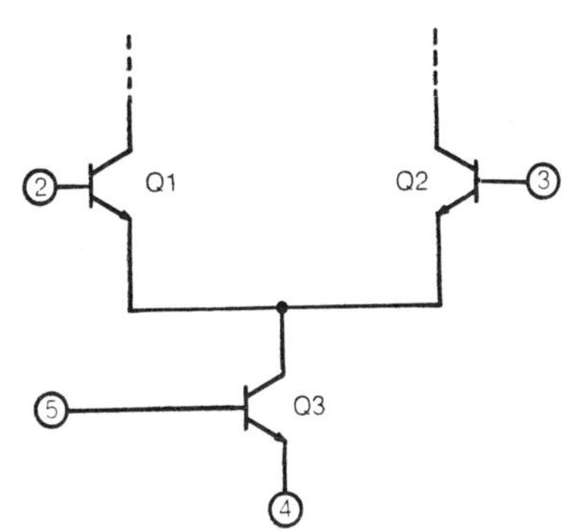

Fig. 3-14. The input stage of RCA's CA3094 revisited. The bias on pin 5 controls the current flowing in the input transistors.

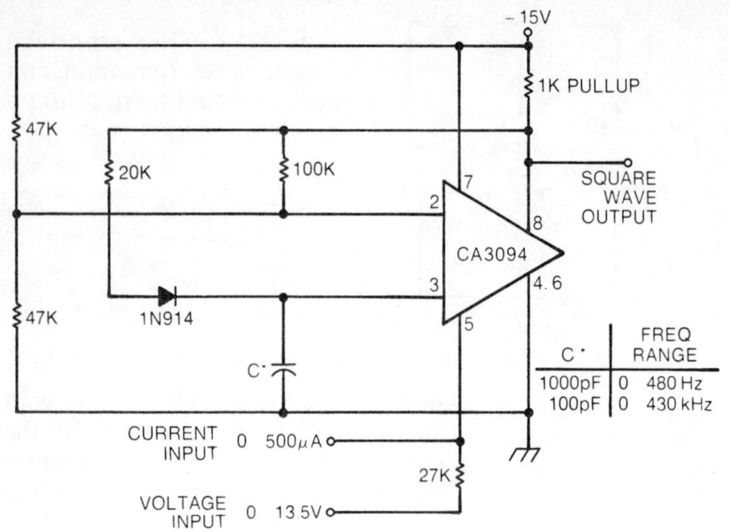

Fig. 3-15. Voltage-controlled oscillator using the CA3094. The bias current into pin 5 controls the rate at which C discharges.

THE LM101A and LM107

Two nearly identical linear IC op-amps are National Semiconductor's LM101A and LM107 (Fig. 3-16). The only difference between the two is that the LM107 incorporates built-in compensation (C2 and C3), while the LM101A provides external connections for compensation components. This gives the LM101A some points in versatility and the LM107 some points in external circuit parts count. Prices of the two ICs are about the same. Both are offered in an 8-pin TO-5 can, a 10-pin flat-pack, and a 14-pin DIP. Maximum supply potential is ±22V. Typical is ±15V.

Function Generator

The function generator in Fig. 3-17 is shown with both an LM101A and an LM107. The LM101A is not frequency compensated, since it is used as a comparator. Another LM107 could as easily have been used in its place; however, if the LM107 were to be replaced by an LM101A, that IC would have to be compensated, because the device is operated in its linear region.

The function generator uses the LM101A as a threshold detector to produce a square wave, which is then integrated by the LM107. Without dwelling on the mathematical implications of integration, here's what happens: Assume that when power

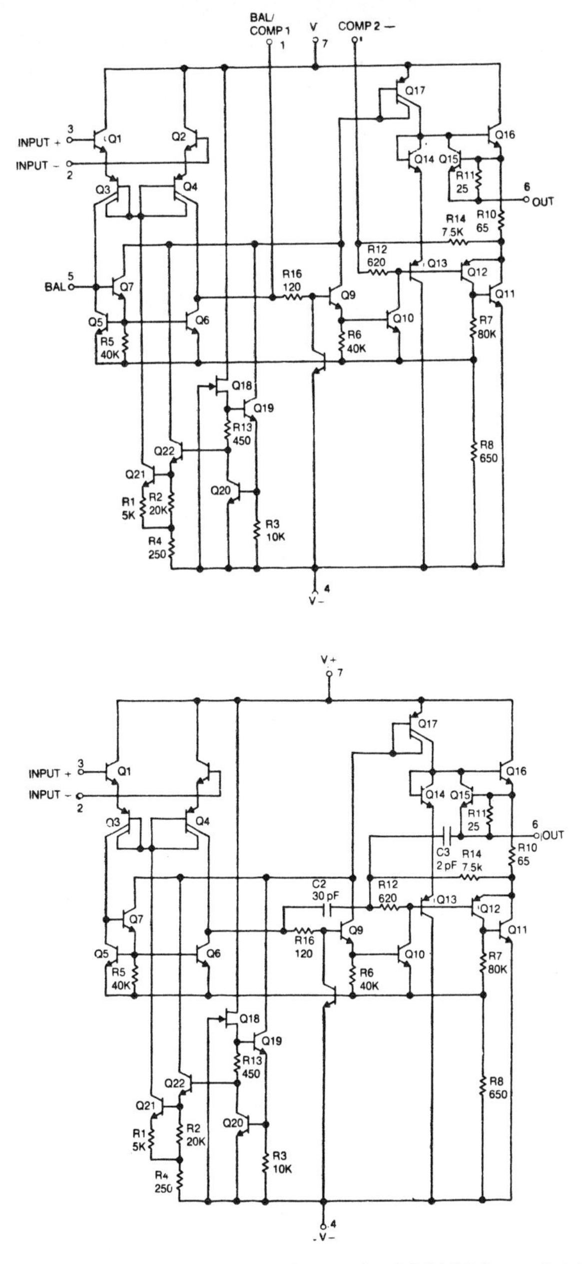

Fig. 3-16. National Semiconductor's LM101A and LM107 have nearly identical circuits. The LM107 has its frequency compensation built-in, though.

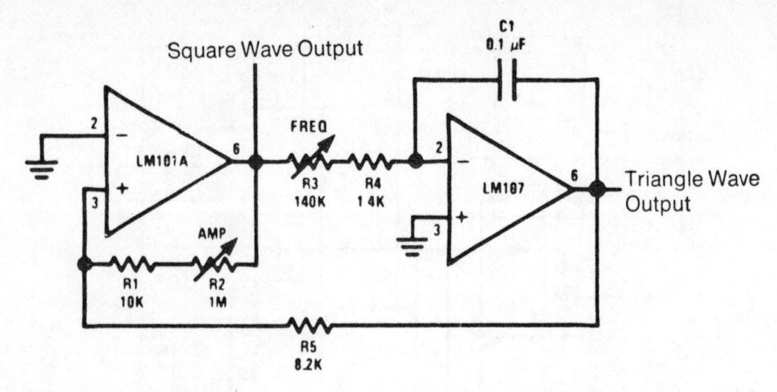

Fig. 3-17. Function generator produces both square- and triangle-wave outputs.

is first applied, the output of the LM101A is immediately driven to the high ($+V_{cc}$) state by the positive feedback. The positive voltage at pin 6 begins to charge capacitor C1 through the combination of R3 and R4. As the voltage at the inverting input to the LM107 rises, the output voltage of that amplifier becomes more and more negative, tending towards $-V_{cc}$. This has two effects. First, it slows down and linearizes the charging current into C1; and second, it reduces the positive

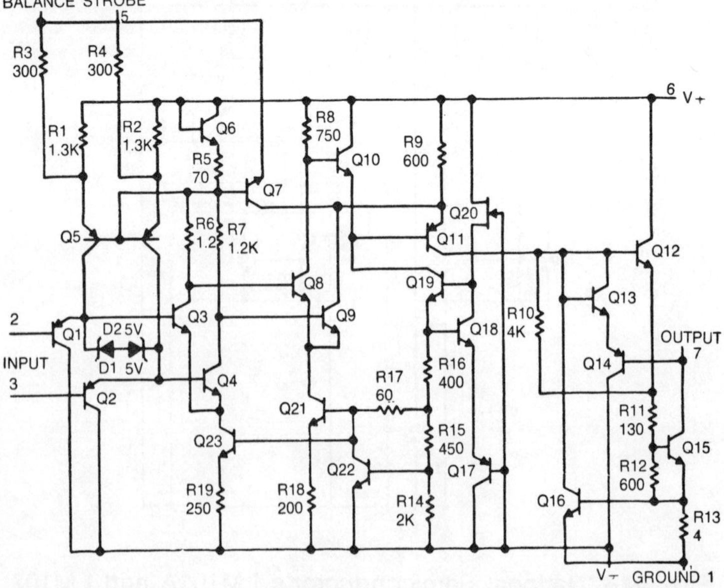

Fig. 3-18. Circuit complexity allows National Semiconductor LM111 comparator to operate with supply voltages as low as 5V.

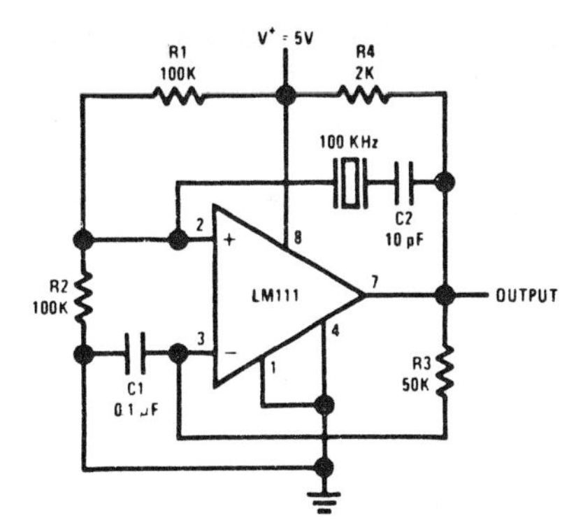

Fig. 3-19. The 10 pF capacitor in this otherwise standard crystal-controlled square-wave generator helps prevent loading down of crystal by circuit connected to output.

voltage on the noninverting input to the LM101A. At a point determined by the setting of R2, the output of the LM101A suddenly changes state; C1 begins to discharge and then charge in the opposite polarity, causing the output voltage of the LM107 to rise toward $+V_{cc}$ at a rate again determined by R3 and R4. This process of square- and triangle-wave oscillation continues as long as power is applied to the circuit.

THE LM111

The National Semiconductor LM111 is a comparator with a more complex circuit than others we have seen in this chapter. Its circuit is shown in Fig. 3-18. The additional complexity al-

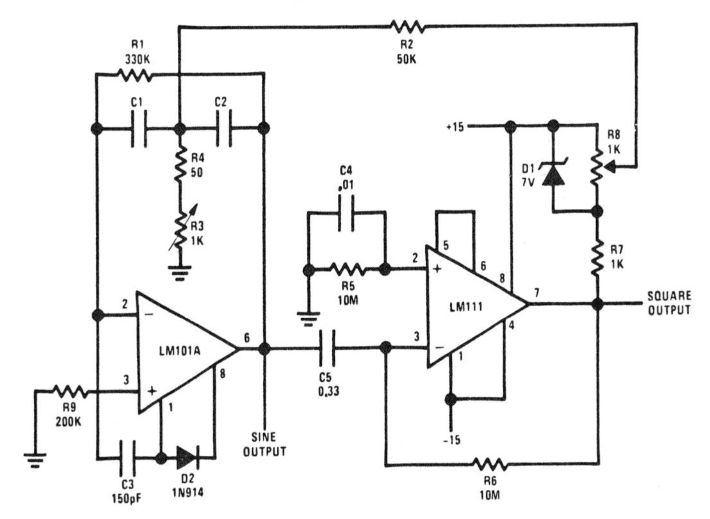

Fig. 3-20. Sine-wave generator operates by filtering square wave.

lows the LM111 to operate from supply voltages as low as 5V. This is useful when the device is required to drive digital logic. Either 8-pin TO-5 cans, 10-pin flat-packs, or 14-pin DIPs may house the LM111 chip.

In the circuit, emitter followers buffer a difference amplifier input stage. This is followed by another difference amplifier consisting of transistors Q8 and Q9, which drives level-shifter transistor Q11. The output transistor is Q15, its collector available at an external pin.

Crystal Oscillator

Figure 3-19 shows a crystal oscillator using the LM111. The external circuit is identical to the circuit in Fig. 3-5, with the addition of capacitor C2, which helps keep the crystal from being overloaded when the circuit drives a capacitive load.

Sine-Wave Generators

A more interesting signal generator using the LM111 and an LM101A is shown in Fig. 3-20. The operation of the circuit is similar to the operation of the oscillators in the preceding example. Instead of a crystal in the feedback loop, however, there is an active filter employing an LM101A. The advantages of using the active filter are that there is a sine-wave output available at the output of the LM101A and that the frequency of such a generator can be much lower than would be possible with a crystal or ceramic filter.

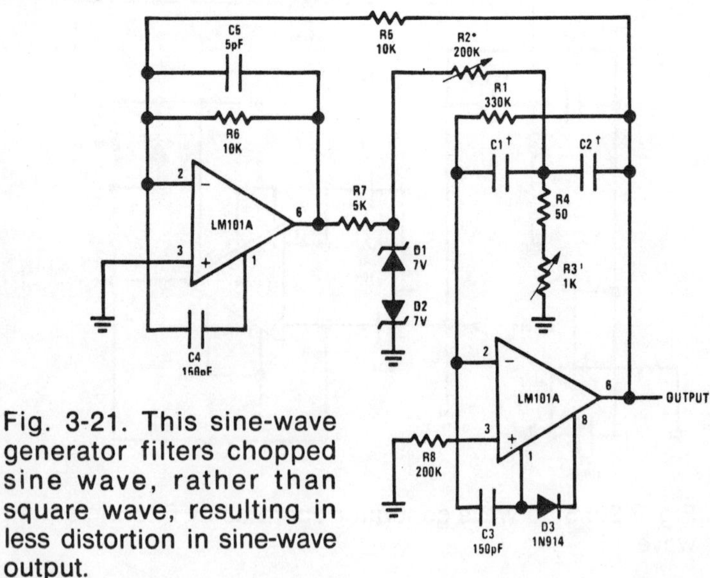

Fig. 3-21. This sine-wave generator filters chopped sine wave, rather than square wave, resulting in less distortion in sine-wave output.

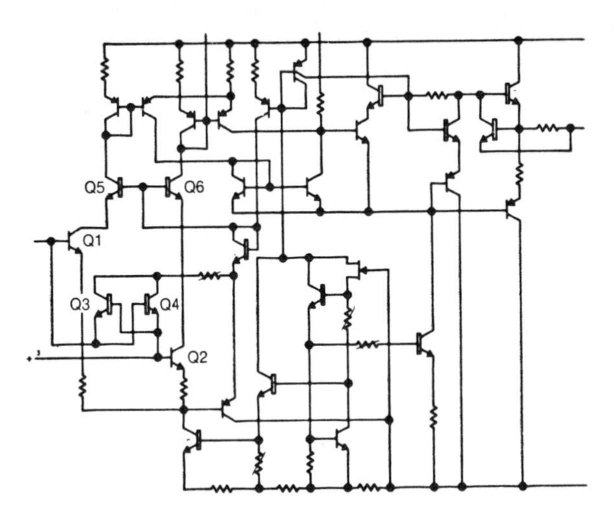

Fig. 3-22. Base width distinguishes between ordinary transistors (flat bases) and super-high-gain ($\beta \geq 5000$) transistors in this circuit diagram of National Semicondutoor's LM108 op-amp.

A sine wave with lower distortion can be obtained with the circuit in Fig. 3-21. The filter remains the same as in the previous example, although moved to the right-hand side of the drawing. It is driven, in this case, not by a square wave, but by a sine wave that is clipped by the back-to-back zener diodes.

Table 3-1 lists values for C1 and C2 that will give various frequency ranges for the above circuits.

The LM108

The National Semiconductor LM108 is a superior operational amplifier with input current levels below even those of FET op-amps. This performance comes about through the use of superhigh-gain transistors (h_{FE} of 5000 or better). In the circuit diagram of Fig. 3-22, these supergain transistors are shown with narrow bases, while the conventional transistors

Table 3-1. Capacitance and Frequency Table for LM108 O-amp Shown in Fig. 3-22.

C1, C2	f_{min}	f_{max}
0.47	18 Hz	80 Hz
0.1	80	380
0.022	380	1.7 kHz
0.0047	1.7 kHz	8
0.002	4.4	20

on the chip are drawn with wide bases. The LM108 is designed to be operated with dual supply voltages up to ±20V. Typical operation is with ±15V. Three package options are available: 8-pin TO-5 can, 10-pin flat-pack, and 14-pin DIP.

The LM108 input circuit consists of supergain transistors Q1 and Q2. These operate in cascode with conventional transistors Q5 and Q6 to drive a level-shifting second stage, which in turn drives a complementary class B output.

Staircase Generator

An interesting circuit that makes use of the high-input impedance of the LM108 is the staircase generator of Fig. 3-23. Staircase generators are used either to transform a serial digital word into an analog output or, as is the case of the circuit in the figure, to generate a function that increases stepwise as a function of time. Such a circuit would be useful as part of a transistor curve tracer (Fig. 3-24). In that application, the staircase generator provides a new bias level for the transistor under test each time it is triggered by the clock signal. Simultaneously, the clock drives the ramp generator that applies V_{cc} to the transistor and drives the scope trace in the horizontal direction.

In Fig. 3-23, the LM101A is used as a monostable multivibrator. The clock input pulses produce a series of positive square wave pulses, each about 32 msec long. During

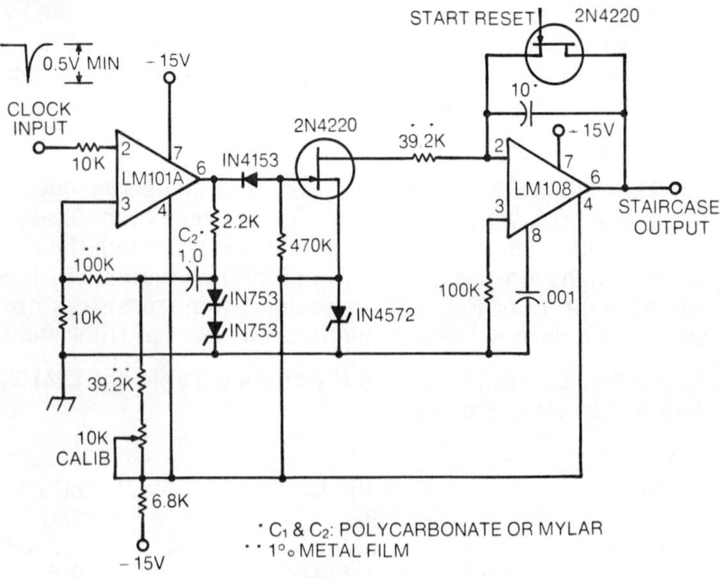

Fig. 3-23. Staircase generator.

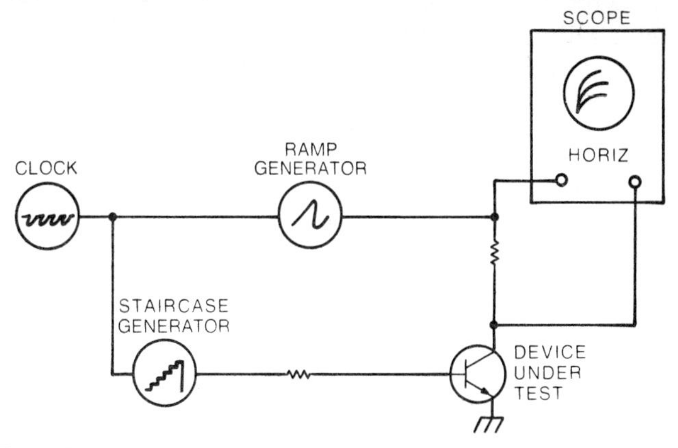

Fig. 3-24. An application for a staircase generator—a transistor curve tracer.

the *on* portion of each pulse, Q1 turns on, applying a regulated −6.4V to the inverting input of the LM108 integrator. The output voltage of the LM108 increases in 0.5V steps during each 32 msec period. During the *off* period, the output voltage is held at a step level by the voltage across C1. When the generator is saturated, or when it is desired to reset the output voltage to zero, the FET across C1 is turned on, discharging the capacitor.

THE LM3900

National Semiconductor's LM3900 is a radical departure in op-amp design, as Fig. 3-25 suggests by its simplicity. The figure represents one of the four amplifiers on the LM3900 chip. The unique aspect of the LM3900 design is its "current mirror" input. The lower transistor and the diode in the drawing form the current mirror. The characteristics of diode and transistor are so matched that collector current exactly "mirrors" base current. Thus, a positive current into the noninverting terminal has the identical effect that a negative current out of the inverting terminal would have. If we place large resistors in series with both inputs, the amplifier becomes voltage, rather than current, sensitive; and the effect is the same as if we had a standard-difference amplifier input. The second transistor in the circuit is a common-emitter gain stage, and this, in turn, drives an emitter-follower output stage that produces the required low impedance output. The LM3900 operates with single-voltage supplies of 4−36V. Typical supply potential is 15V. A 14-pin DIP is standard for the IC.

National has coined a new circuit symbol for its LM3900 *Norton* amplifier, to emphasize its current input. This symbol

Fig. 3-25. National Semiconductor's LM3900 "Norton" amplifier features a current mirror input, explained in the text. If large external resistors are placed in series with the inputs, the Norton amplifier behaves much like a conventional op-amp. There are four Norton amplifiers on each LM3900 chip.

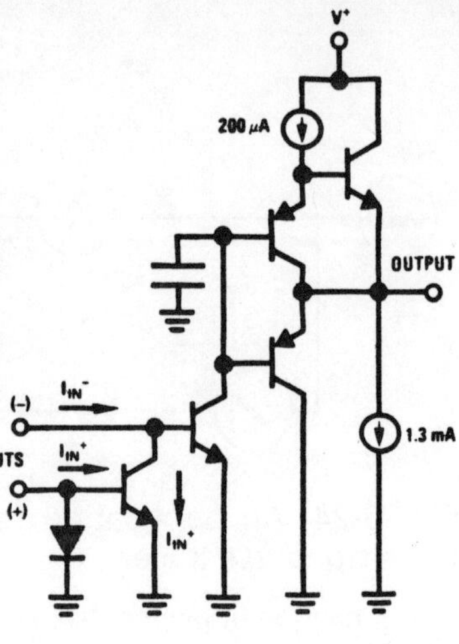

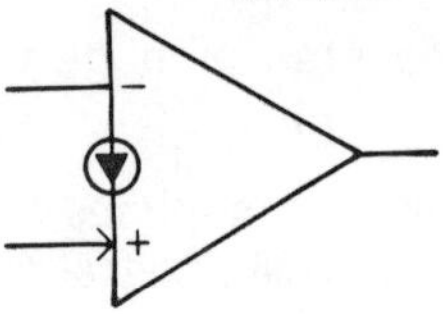

is shown in the figure. The "Norton" referred to is the Bell Laboratories researcher who pioneered the use of the current analog to the Thevenin equivalent circuit.

Another Staircase Generator

As an application of the LM3900, consider the staircase generator of Fig. 3-26. The first stage is a pulse generator that makes use of the LM3900's ability to pass current through its inputs. The amplifier is set for high gain by the positive feedback provided by R1 and R2. Capacitor C1 charges through R4 until its voltage exceeds the voltage at pin 1. Then the output of

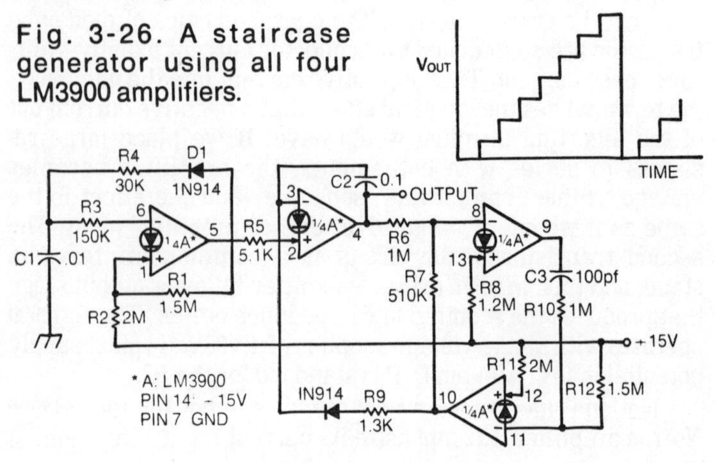

Fig. 3-26. A staircase generator using all four LM3900 amplifiers.

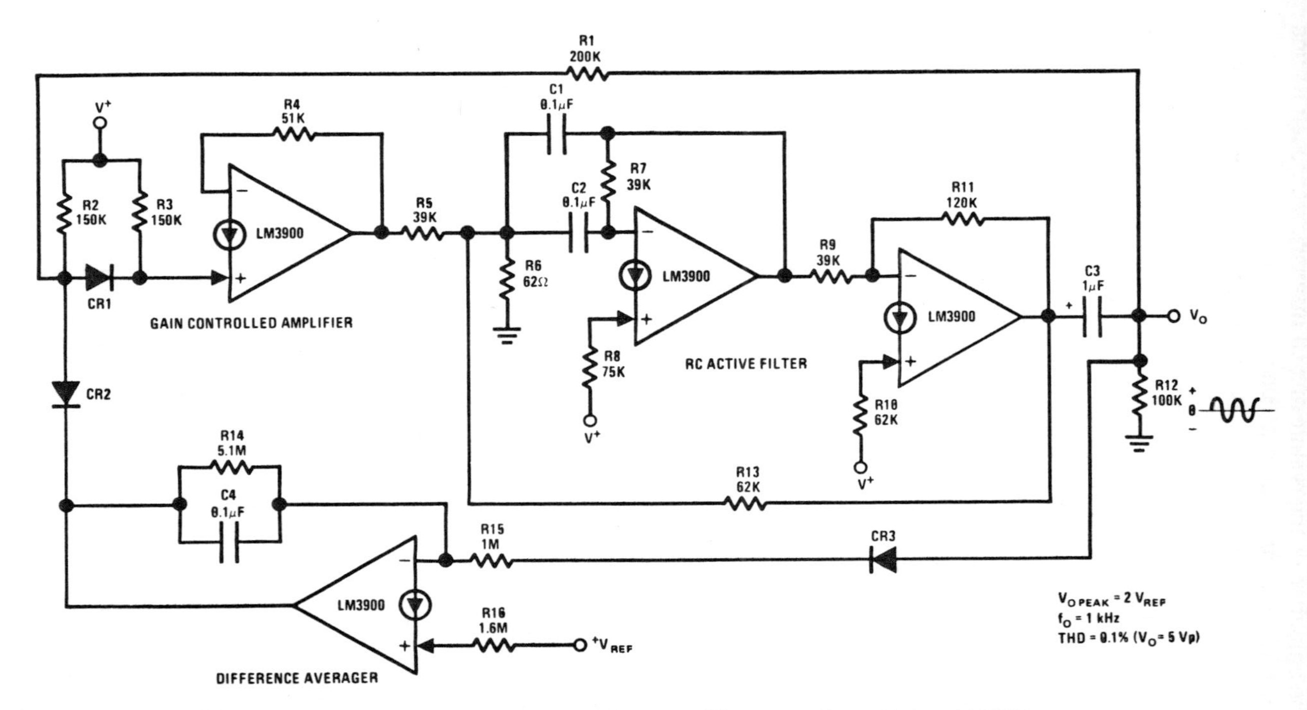

Fig. 3-27. A precision sine-wave oscillator using the LM3900.

the amplifier changes state and C1 discharges through R3 and the LM3900 input stage to ground.

The second amplifier develops the staircase output in much the same manner that the LM108 in the previous staircase generator did. During each positive pulse, the output of the amplifier rises rapidly to a new level. As soon as the pulse is cut off, the amplifier holds at the new voltage, due to the action of C2.

The next two amps provide an automatic reset signal when the staircase output rises to within 80% of the supply voltage. The third amp samples the output of the staircase generator; when the output reaches the proper level, it turns on, causing the last amplifier to send a 100 μsec reset pulse (controlled by the values of C3 and R10) to the second stage, resetting its output to zero.

Precision Sine-Wave Oscillator

Another project using the LM3900 is shown in Fig. 3-27. This is a sine-wave oscillator with peak-to-peak output precisely held to twice some reference voltage.

The heart of the oscillator is the two-amplifier RC active filter. The input stage (the middle amplifier) does the actual filtering. The second stage provides feedback gain that produces a very high *Q*. The output of the filter is fed back around the circuit through R1 to the gain-controlled amplifier. This introduces enough gain for instability, and oscillation results. The gain of this amplifier is controlled by a signal from the fourth amplifier, the difference averager, which compares the output level to a reference voltage.

Phase-Locked-Loop Frequency Synthesizer

As a final signal-generator project, consider the phase-locked-loop frequency synthesizer in Fig. 3-28. It uses an MC4344 phase detector with a built-in op-amp for the active filter, an MPS6571 transistor, an MC4324 voltage-controlled oscillator, and two MC4316 programmable dividers. All the ICs are made by Motorola.

The passive components in the filter are selected to give a maximum lockup time of 1 msec when a new frequency is selected. The 100 pF capacitor sets the VCO for a range of 2–3 MHz. The reference frequency f_i is 100 kHz, giving 100 kHz steps between output frequencies, e.g., 2.0, 2.1,... 2.9, 3.0 MHz. This means that the signal from the VCO must be divided by 20, 21,...29, and 30 to match the reference signal. The dividers are programed by entering the correct hexadecimal word into the *units* and *tens* dividers. For example, to divide by 29, the

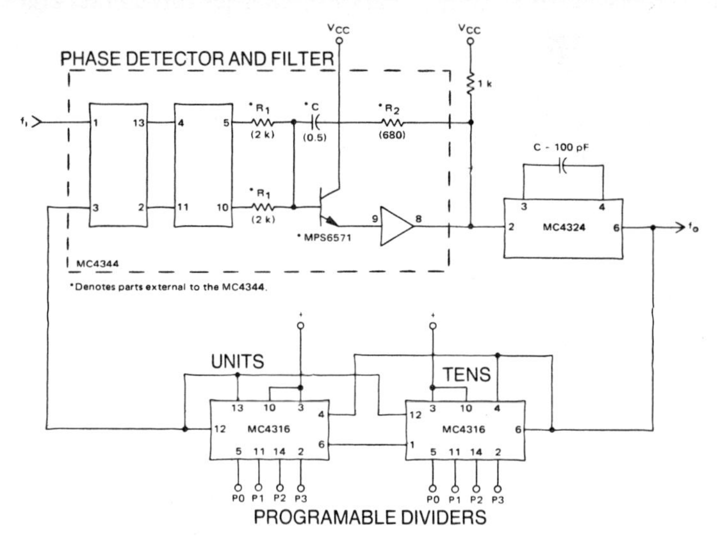

Fig. 3-28. A 2 to 3 MHz PLL frequency synthesizer programmable in 100 kHz frequency steps.

inputs of the *tens* divider would be programed 0010, and the inputs of the *units* divider would be programed 1001. The phase detector would then drive the VCO to a frequency of 29 MHz, since this divided by 29 will produce a 100 kHz signal to match the input from the reference generator.

RF—IF Amplifiers and Detectors

This chapter shows some ways linear ICs can be used in AM, FM, SSB, and CW receivers. Circuit complexity varies from very simple to full LSI (large-scale integration).

THE CA3005/CA3006

Both the RCA CA3005 and CA3006 share the same circuit configuration (Fig. 4-1) and 12-pin TO-5 package. The CA3006 has a lower offset voltage requirement than the CA3005, 1 mV vs 5 mV maximum. In contrast to many of the ICs in the preceding chapters, these are RF devices, usable up to frequencies of 100 MHz.

If you have read and understood Chapter 1, you will find no need of a detailed explanation of the circuit diagram. It is worth noting that virtually every circuit node is connected to an external pin. This permits considerable latitude in biasing and operating the IC. For example, pins *4* or *5* (or both *4 and 5*) can be shorted to pin *8* to vary the bias and, consequently, the gain of the amplifier; or the gain can be controlled by an

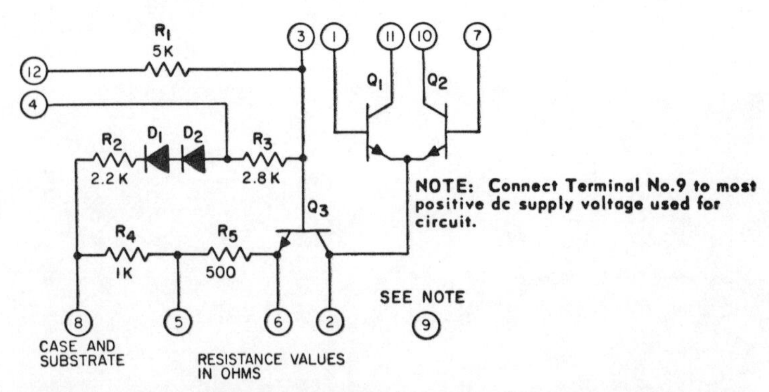

Fig. 4-1. The CA3005/CA3006 is a difference amplifier usable up to 100 MHz. Its internal nodes are quite accessible, allowing a variety of bias options. It can even be used in cascode configuration with a signal fed into pin 3 and one of the difference amplifier transistors disabled.

external bias applied to pin *3*, as was done in the last chapter. It is even possible to disable one or the other of the difference amplifier transistors, inject a signal into pin *3*, and operate the amplifier in a cascode configuration for high gain at VHF.

Product Detector

A product detector is actually a mixer used to detect SSB or CW signals. The carrier is regenerated by means of a beat frequency oscillator (BFO) and mixed with the sideband or CW signal from a receiver's IF stage. The filtered output of the detector is an audio signal representing either the information originally impressed on the transmitted sidebands or a tone for CW copy. The product detector in Fig. 4-2 uses a CA3005/CA3006. The double-ended modulated signal from the level-shifting transformer is coupled to the difference amplifier inputs, and the regenerated carrier (BFO signal), which must have an amplitude approximately ten times that of the modulated signal, is injected into pin *3*. The double-ended output is used to drive the receiver's audio amplfier. The network consisting of the two 1K resistors and two 0.02 μF capacitors is a filter that bypasses RF from the detector output.

THE LM118 AND LH0062

Two high-speed, wide-bandwidth op-amps from National Semiconductor are shown in Figs. 4-3 and 4-4. The LM118 and LH0062 circuits are identical except for their input stages. The

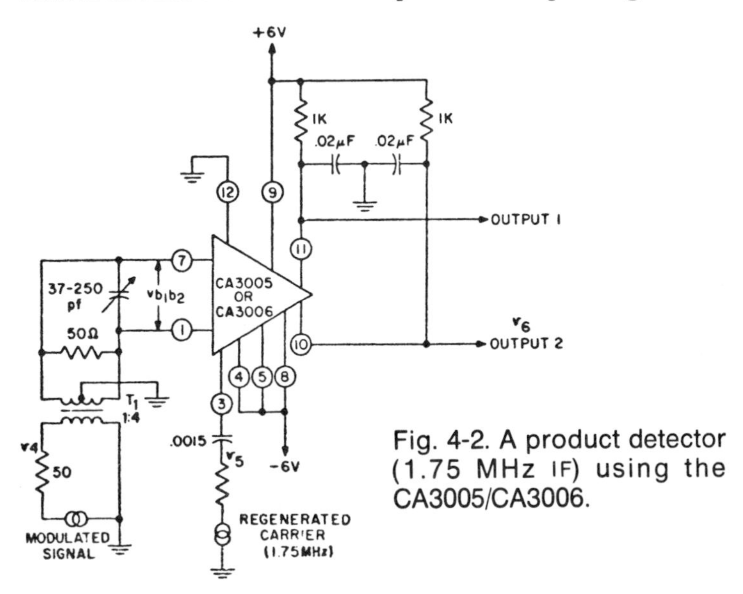

Fig. 4-2. A product detector (1.75 MHz IF) using the CA3005/CA3006.

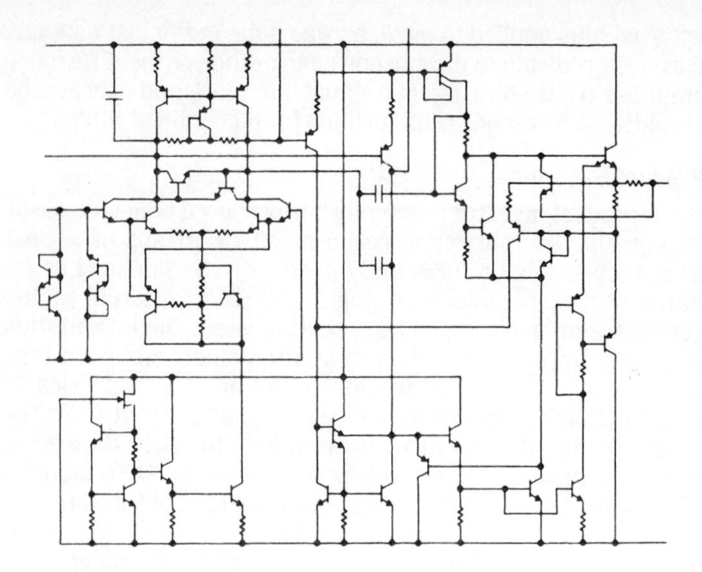

Fig. 4-3. The LM118 high-speed, wide bandwidth op-amp uses darlington connected bipolar transistors in its input circuit. Voltage gain is around 200, 000 (100 dB).

LM118 uses bipolar transistors in darlington pairs; the LH0062 uses field-effect transistors. The FET version, naturally, has the lower input bias current by a factor of 10^5, but its voltage gain is only half that of the bipolar IC. Of course, with gains on the order of 100 dB, six decibels here or there isn't likely to be missed. Typical slew rate for both units is around 70V/μsec, and bandwidth is 15 MHz.

The LM118 and the LH0062 are both available in 8-pin TO-5 cans, 10-pin flat-packs, and 14-pin DIPs. They are designed to operate from dual-voltage supplies of potentials up to ±20V. Typical potential is ±15V.

High-Speed Peak Detector

Both the LM118 and the LH0062 are used to advantage in the very high-speed peak detector of Fig. 4-5. This is a positive peak detector. The diodes could be reversed if a negative peak detector were desired. The 1N914 negative feedback loop reduces the gain of that amplifier effectively to unity during negative portions of the input cycle. During positive portions of the cycle, gain is set by R2. The 2N930 transistor is used for the low leakage of its base-to-collector junction. The LH0062 serves as a voltage follower, keeping the output from loading

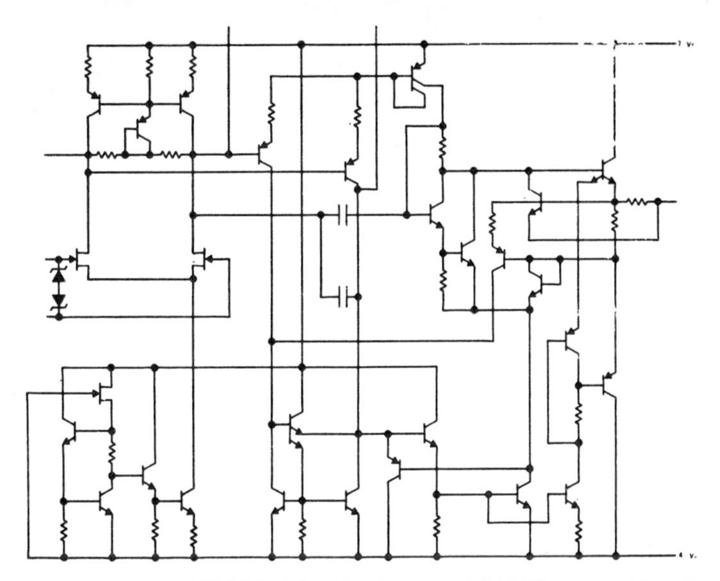

Fig. 4-4. The LH0062 is identical to the LM118 except for its FET input stage. Its voltage gain is only 100, 000 (94 dB), but its input resistance is extremely high.

down C2. The small trimmer capacitor, C1, is set to cancel stray board capacitance. This detector will acquire a 10V signal in under 4 μsec.

THE CA3028

The RCA CA3028 (Fig. 4-6) is another simple circuit with applications up to 120 MHz. It is much like the CA3005 and it can be used in a similar fashion.

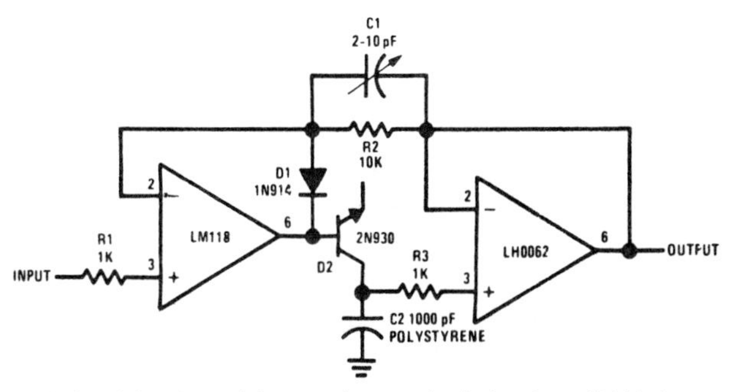

Fig. 4-5. Is this the ultimate in peak detectors? We've come a long way from the galena crystal and cat's whisker, but the principle remains the same.

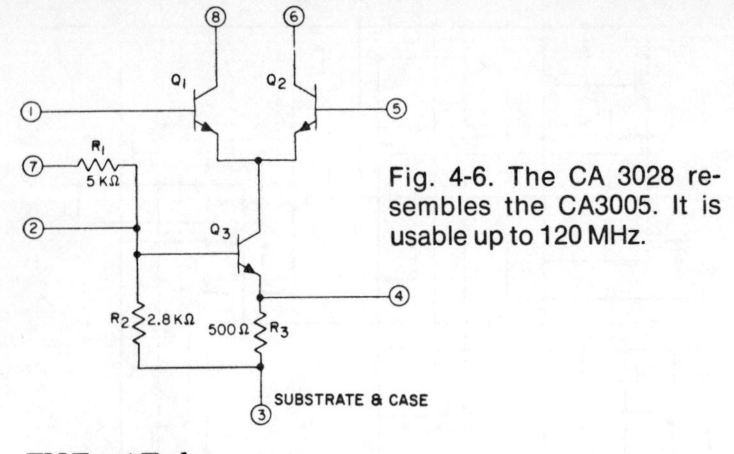

Fig. 4-6. The CA 3028 resembles the CA3005. It is usable up to 120 MHz.

FM Front End

An FM front end for the 88–108 MHz broadcast band is shown in Fig. 4-7. Its noise figure is slightly higher than that of a typical discrete component front end, but its parts count is less and its sensitivity is reasonably good.

The RF stage is conventional. The 1 μH RF choke between the input terminals of the difference amplifier keeps the two transistors at the same DC base potential. The amplifier is single-ended in both input and output; pins *5* and *8* are RF bypassed to ground.

The signal from the output tank circuit of the RF amplifier is fed to the gain control transistor of the CA30028 converter. The differential stage is used as the local oscillator. To keep the drawing reasonably simple, pin *1* is shown twice. The components connected to pin *1* are for "housekeeping"; pins *5* and *8* are connected for oscillator feedback.

THE LM172

National Semiconductor's LM172 (Fig. 4-8) bears some resemblance to the LM170. There is the same kind of AGC stage in the input and the same kind of cascade gain stage. Missing are the squelch and external AGC inputs; added is an active peak detector.

Operation is easy to understand. Capacitors C1 through C4 are external to the IC. Input is applied to emitter follower Q2. This stage feeds the gain stage consisting of Q6, Q7, and Q8 in cascade. The detector stage is a difference amplifier with a level-shifting output and a diode detector in the feedback loop. Part of the output is fed back to Q3 in the input stage, where it performs its AGC function by robbing current from Q2.

Fig. 4-7. A front end for a broadcast FM receiver. It looks complicated with all the lines, but there are surprisingly few parts.

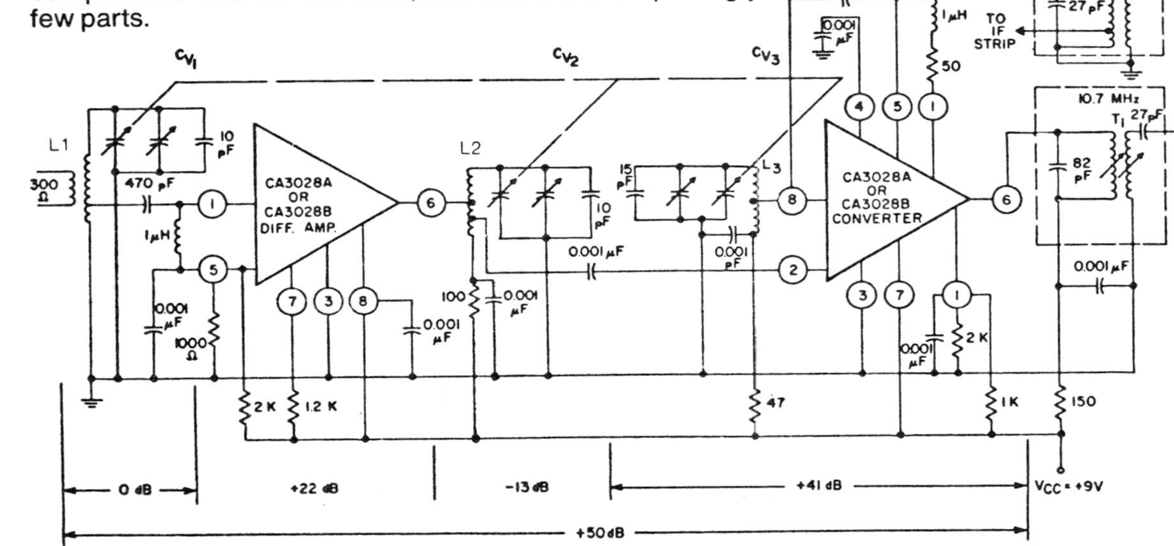

L1: 3-3/4 Turns of awg 18 tinned copper wire. 5/16 in. winding length on 9/32 in. diameter form: tapped at 1-3/4 turns. Primary consists of 2 turns of awg 30 silicon enamelled copper wire.

L2: 3-3/4 turns of awg tinned copper wire. 5/16 in. winding length on 9/32 in. diameter form: tapped at 2-1/4 turns and at 3/4 turn.

L3: 3-1/2 turns of awg 18 tinned copper wire. 5/16 in. winding length on 9/32 in. diameter form. tapped at 2 turns.

Cv: Ganged 15 pF variable capacitors. with trimmers 1 – 3.

T1: Mixer transformer. TRW 22484. or equivalent.

T2: Mixer transformer. TRW 22485. or equivalent.

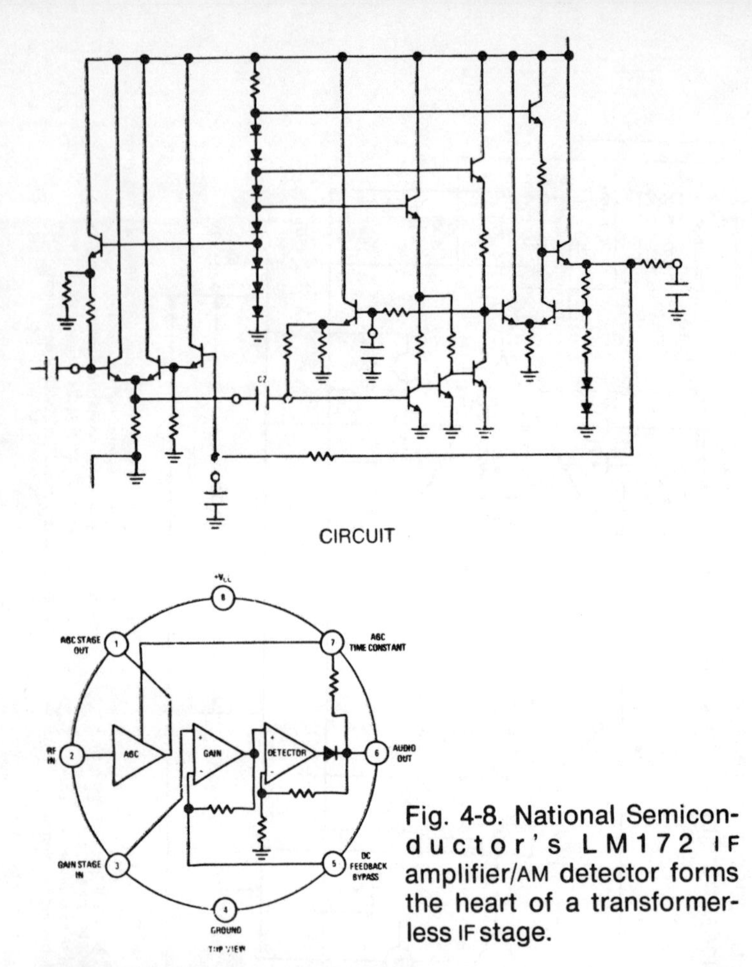

Fig. 4-8. National Semiconductor's LM172 IF amplifier/AM detector forms the heart of a transformerless IF stage.

Broadcast Receiver

The LM172 is mainly intended as an IF amplifier/AM detector. However, as Fig. 4-9 shows, it can be used for a simple, one-IC tuned-RF (TRF) broadcast band receiver as well. The circuit is simple and easy to follow by reference to Fig. 4-8. The only factor worthy of special note is that the DC bias for the 2N2219 output transistor is obtained directly from pin 6 of the IC, eliminating the need for at least two resistors.

Since the gain of the LM172 falls off rapidly above 2 MHz, modifications of this circuit will not be suitable for shortwave reception unless they incorporate a local oscillator and a converter. For frequencies below the broadcast band,

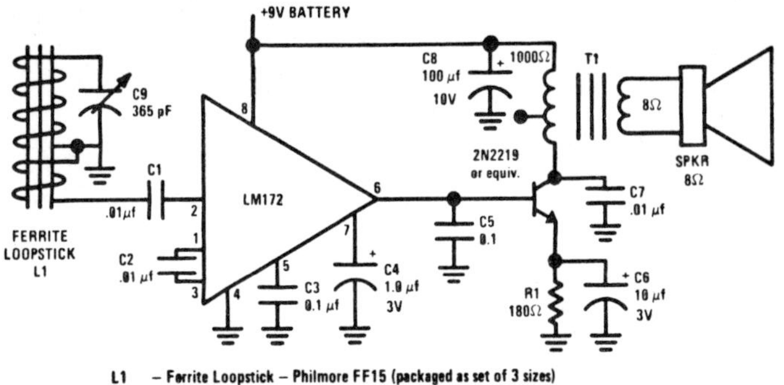

L1 – Ferrite Loopstick – Philmore FF15 (packaged as set of 3 sizes)
C9 – Sub-miniature variable capacitor – Philmore 1949G – 365 pF max.
T1 – Midget Audio Transformer, 1000Ω:8Ω – Archer 273-1380 (Radio Shack, Inc.)
SPKR – 2" PM Speaker, 8Ω, 0.1 watt – Philmore TS20

Fig. 4-9. A little low on selectivity, this TRF broadcast receiver is still an impressive demonstratration of a one-IC receiver.

however,the simple TRF circuit with an RF preamplifier should be able to receive long-wave navigation beacons and other services in that part of the spectrum.

THE LM273/274

A pair of more complex IF amplifiers and detectors is National Semiconductor's LM273 and 274. Both devices have a common block diagram, shown in Fig. 4-10. The difference between the two is in the output impedance of their respective first sections. The LM273 is optimized to drive low-impedance loads, such as mechanical or ceramic filters. The LM274 has a

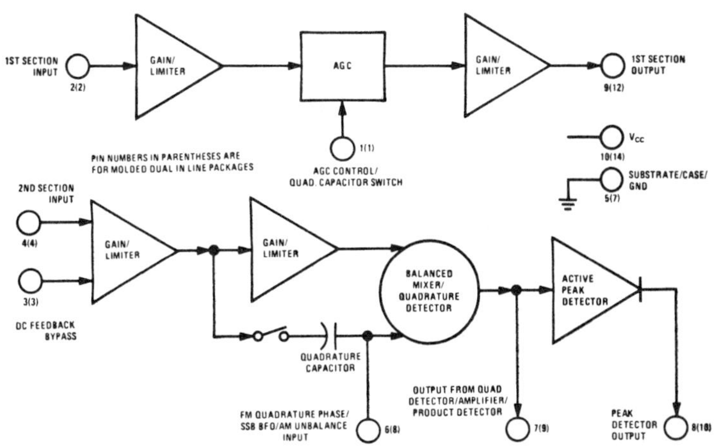

Fig. 4-10. There are two sections to the LM273/274 IF amplifier/AM, FM, SSB detector. Grounding the AGC input closes the quadrature capacitor switch.

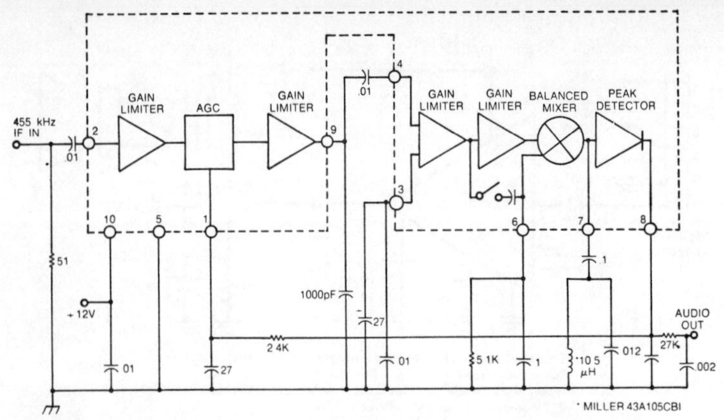

Fig. 4-11. The LM273 used as an AM detector. With the AGC input ungrounded, the quadrature capacitor switch is open. The 5.1K resistor disables the balanced mixer. The LC network connected to pin 7 improves signal-to-noise ratio.

high output impedance for driving high-impedance crystal, ceramic, or LC filters.

The LM273/274 can be used for AM, SSB, or FM detection. For FM detection, the AGC input pin is grounded. This closes the switch inside the IC that connects the quadrature capacitor to the input circuit of the balanced mixer. If an external phase-shift network is connected to the balanced mixer, the mixer output will be a pulse-duration-modulated signal that

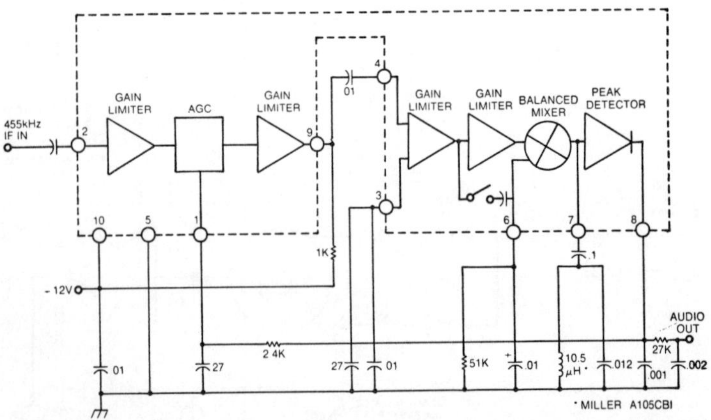

Fig. 4-12. The LM274 used as an AM detector. This is the same as the previous circuit, except for the filter connecting the two sections of the IC.

can be filtered and coupled to an audio amplification stage, bypassing the peak detector. In the FM mode, the gain amplifiers in the IC function as limiters.

For AM detection, the balanced mixer is unbalanced by connecting a resistor between its input and ground, and the output of the active peak detector is fed back to the AGC stage through an RC network. This opens the quadrature capacitor switch.

For SSB/CW operation, the balanced mixer is used as a product detector by injecting a BFO signal into its input. AGC feedback can be used for SSB, or a manual gain control may be used for CW.

AM, FM, and SSB/CW Detectors

Figures 4-11 through 4-16 show detailed circuits for the LM273/274 used for AM detection, wide- and narrow-band FM detection, and SSB/CW.

THE CA3123

The CA3123 (Fig. 4-17) is an RCA approach to a complete AM-receiver IC. It was developed for automobile radios. It lacks only a detector, which for this purpose can be a simple diode and capacitor and an audio output stage, for which any of the appropriate circuits in Chapter 2, among others, would suffice.

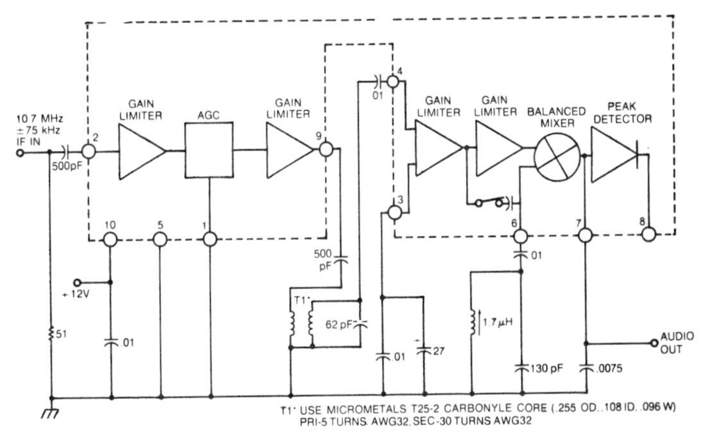

Fig. 4-13. LM273 connected for wideband FM. The AGC input is shorted to ground, closing the quadrature capacitor switch. The capacitor, along with the LC network connected to pin 6, provides a phase shift at this input to the balanced mixer that varies with frequency with respect to the other input to the mixer.

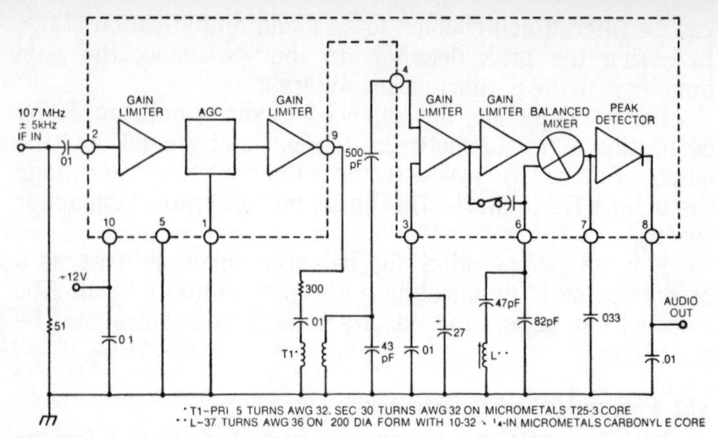

Fig. 4-14. Another LM273, this time with suitable components for narrowband FM. For added audio gain, the output is taken from the peak detector, which in this case functions as an amplifier.

The CA3123 features a cascode RF amplifier, an oscillator/mixer (such as presented earlier in the chapter), and a single-ended difference amplifier (IF amplifier). It is supplied in a 14-pin DIP, and it is designed to operate from a single-output 9V supply.

Superhet Broadcast Receiver

RCA's version of an auto radio receiver using the CA3123 is shown in Fig. 4-18. See Table 4-1 for detailed coil and transformer data. The IF is 262 kHz.

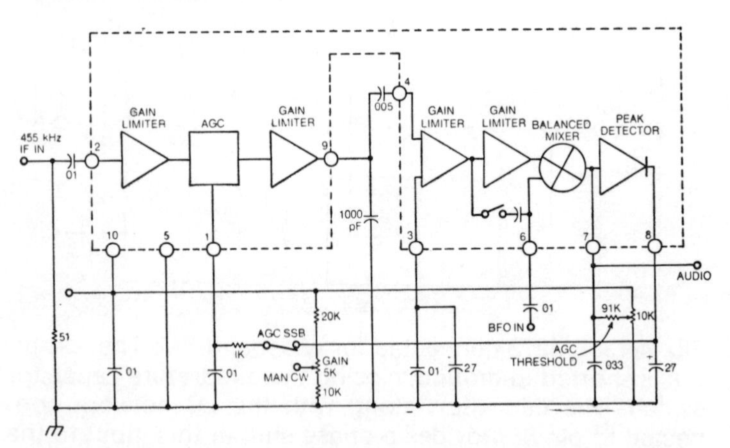

Fig. 4-15. LM273 used for SSB/CW. Balanced mixer is used as a product detector with 455 kHz BFO injected at pin 6.

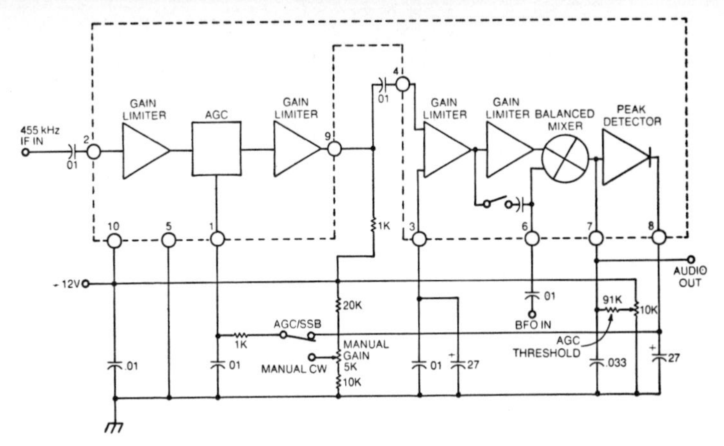

Fig. 4-16. Another SSB/CW detector, this time using the LM274. Note minor change in interstage coupling.

THE 561B

The Signetics 561B is a phase-locked-loop FM and AM detector. For FM detection, it functions as described in Chapter 1. For AM detection, the VCO is locked to the carrier frequency, and its output, shifted 90°, is used as the local oscillator input to a product detector built into the 561B chip. Figure 4-19 indicates 561B pin assignments.

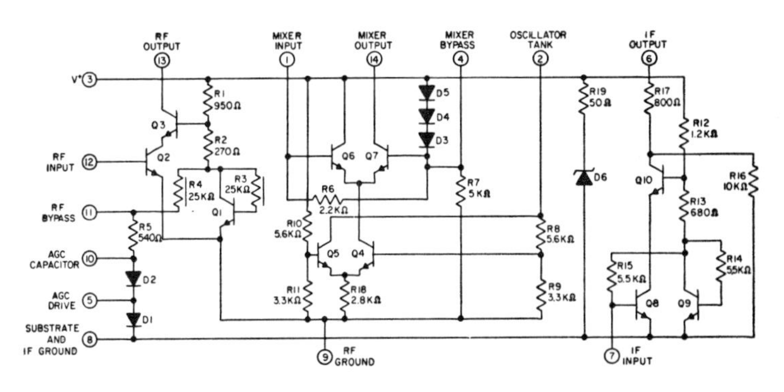

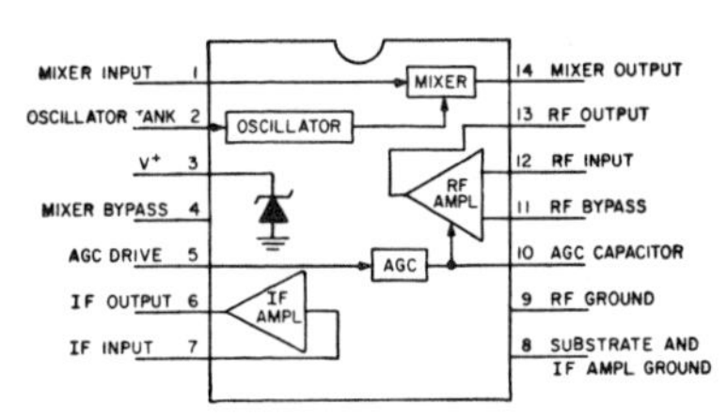

Fig. 4-17. The CA3123 is RCA's version of a single-chip receiver.

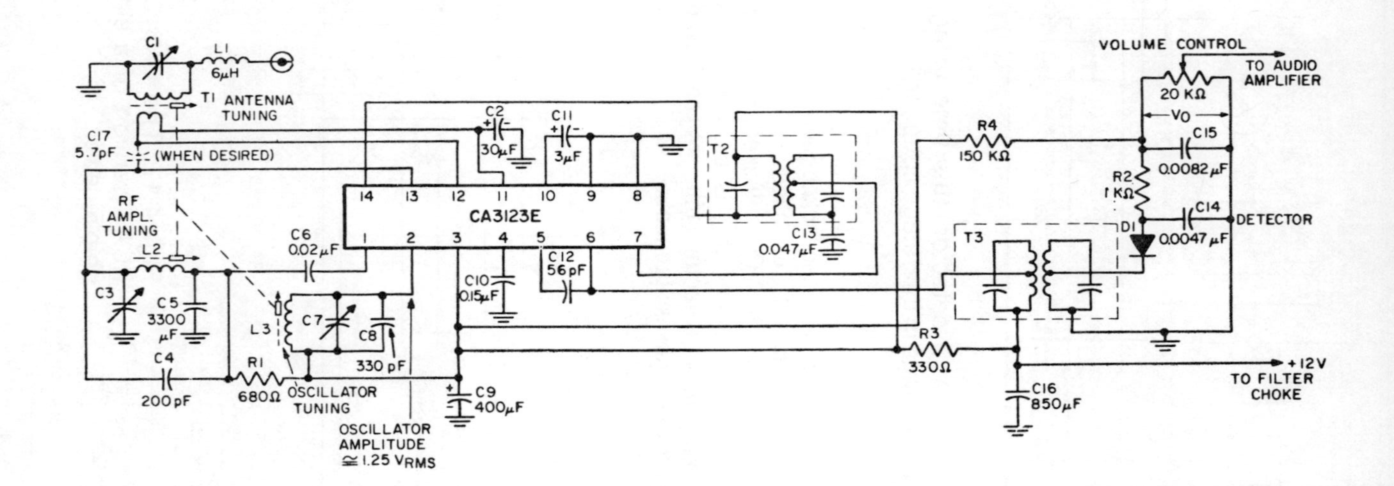

Fig. 4-18. An automobile receiver using the CA3123. Coil and transformer data are given in Table 4-1.

Table 4-1. Coil and Transformer Data for Auto Radio.

Transformer	Symbol	Frequency	Inductance μh (≈)	Capacitance pF (≈)	Q (≈)	Total Turns To Tap Turns Ratio	Coupling
First IF:							
Primary	T_2	262 kHz	2840	130	60	none	
Secondary			2840	130	60	or 30:1 31:1	critical ≈0.017 ≈1/Q
Second IF:							
Primary	T_3	262 kHz	2840	130	60	8.5:1	–
Secondary			2840	130	60	8.5:1	critical ≈0.017 ≈1/Q
Antenna:							
Primary	T_1	1 MHz	195	(C_1)–130	65		
Secondary		Adjusted to an impedance of 75 Ω with primary resonant at 1 MHz. Coupling should be as tight as practical. Wire should be wound around end of coil away from tuning core.					
Coils	L_1	7.9 MHz	6		50		
	L_2	1 MHz	55		50		
	L_3	1.262 MHz	41		40		

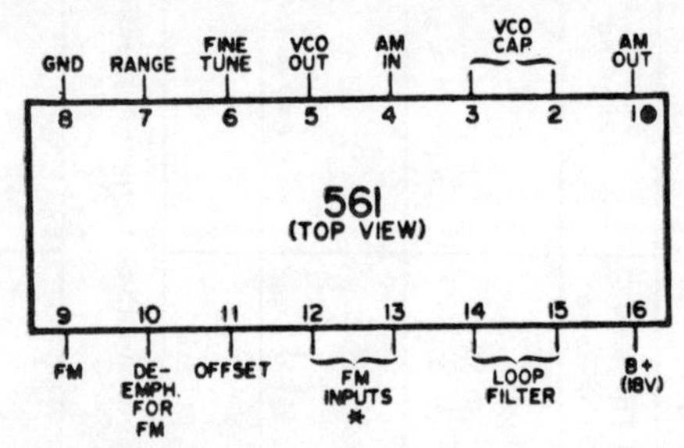

Fig. 4-19. Pin assignments on the Signetics 561B AM/FM receiver.

PLL AM Broadcast Receiver

Figure 4-20 shows the PLL AM receiver circuit. The 480 pF variable capacitor is tuned to set the frequency of the VCO to 540 kHz when the tuning pot slider is at ground. Increasing the current into pin *6* then tunes the receiver across the broadcast band.

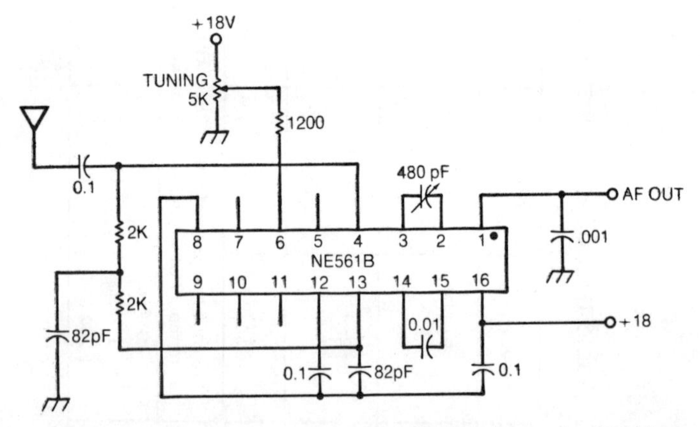

Fig. 4-20. An AM receiver using the 561B phase-locked loop.

Regulators and Power Supplies

There was a time when a regulated power supply was one with one or more OA2 gas regulator tubes and associated resistors across its output. The state of the art has come a long way since then, and voltage regulators have become both more precise and more complex. Complexity, however, need not concern us when we are dealing with linear integrated circuits. What was unthinkable complexity in the days of discrete components is available today in complete regulator ICs costing less than two dollars.

THE μA723

Perhaps the most common linear IC voltage regulator is Fairchild's μA723 and its 723-designated cousins produced by other manufacturers. Prices for 723s presently start around a dollar and a half and will undoubtedly go lower. Since the 723 is representative of all IC voltage regulators, we will take a close look at it in this chapter.

The simplified schematic in Fig. 5-1 will help explain the circuit. In the bias supply, Q1 serves as a simple constant-current source for zener diode D1. The zener voltage is applied to voltage divider R1–R2, and is used to supply the current generators in the rest of the IC.

Zener diode D2 is the basic reference source for the regulator. The 6.2V drop across D2 is added to the base–emitter voltage across Q6 so that the total reference voltage is approximately 7.15V. The emitter–base drop across Q6 is set by current source Q3 so that the negative temperature coefficient ot Q6 exactly cancels the positive temperature coefficient of D2. Thus, the reference voltage is unaffected by temperature.

The regulating device is a difference-type error amplifier. A signal derived from the reference voltage is applied to the noninverting input and a signal derived from the output voltage is applied to the inverting input. The single-ended output of the error amplifier is proportional to the difference between the two inputs. The output stage is a double emitter

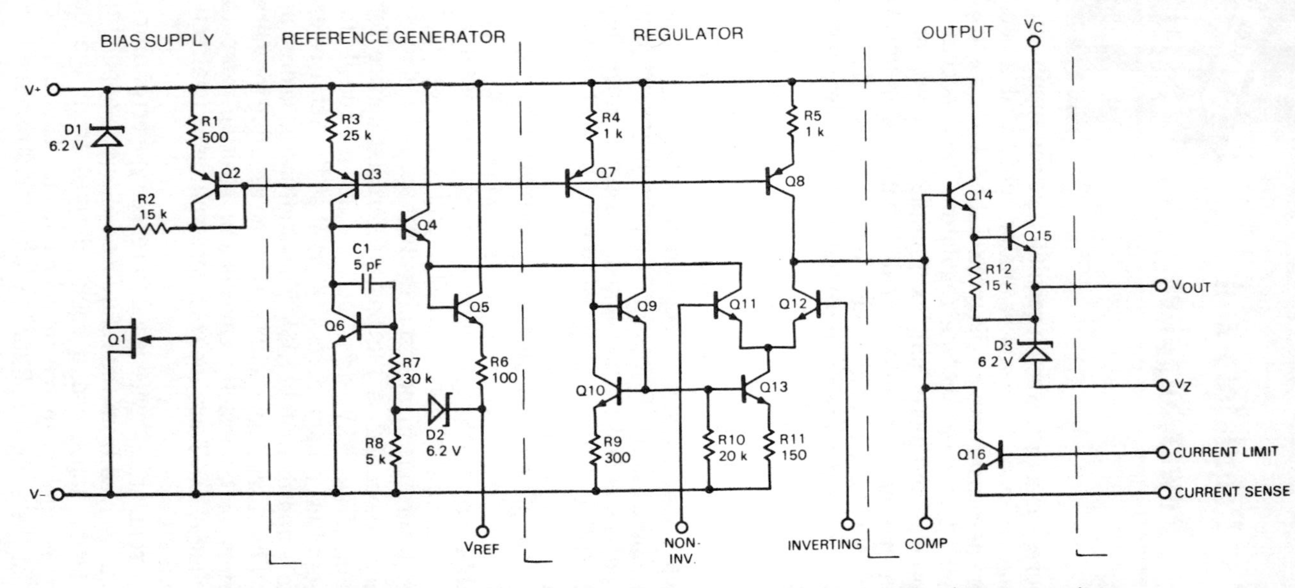

Fig. 5-1. A simplified schematic of the ubiquitous 723-type voltage regulator.

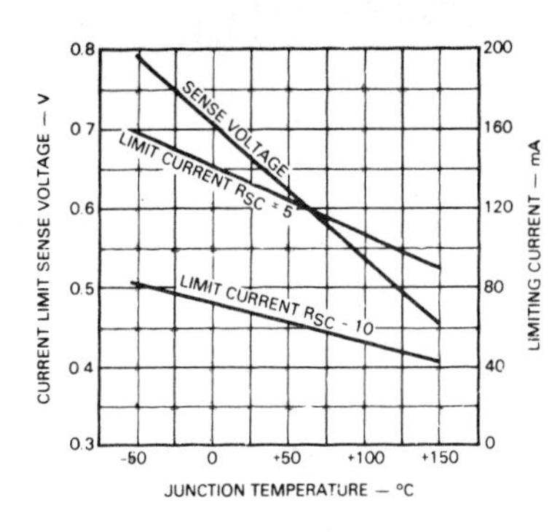

Fig. 5-2. Current limiting performance of the 723 regulator. This graph defines output current as a function of temperature for two values of R_{SC}. Sense voltage is the voltage that must be developed across R_{SC} to make the current-limiting transistor conduct.

follower consisting of Q14 and Q15. Zener diode D3 is used in only a few instances, notably those where the voltage applied to the inverting terminal is very low. Transistor Q16 is used for current limiting. When a voltage greater than a certain level, termed V_{sense}, appears across this transistor's base and emitter terminals, Q16 sinks current from current source Q8, limiting output current. Figure 5-2 identifies V_{sense} and output current at different temperatures for certain values of short-circuit limiting resistance R_{sc}.

The 723 can handle potentials up to 40V between its input terminals. It requires a minimum 3V differential between input and output, and can tolerate up to a 5V differential. The devices are available in 14-pin DIPs and 10-pin TO-5 cans. The DIP is capable of dissipating 1W; the can is rated at 850 mW maximum. Zener diode D3 is not available in the can package.

Basic Positive Regulator—Output Less Than 7.15V

One of two basic positive regulator circuits is shown in Fig. 5-3. In this version, regulated output is limited to 7.15V, the reference voltage available from the IC. Regulated voltages less than this are available by selection of R1, R2, and P. Since the current out of the V_{ref} terminal is limited to 15 mA maximum, the sum of R1, R2, and P must always be greater than 500Ω. Higher values are preferable. Resistor R3 is selected to equal the parallel resistance of R1, R2, and P. It cancels the temperature effects of these resistors. The short-circuit limiting resistor, R_{sc}, may be anywhere in the 0–10Ω range. The more resistance, the more current limiting, as Fig. 5-2 indicates, but also, the less current available.

Capacitor C1 is not always necessary, except where leads from the unregulated supply are long. Zener diodes are relatively noisy devices, and capacitor C2 is desirable in

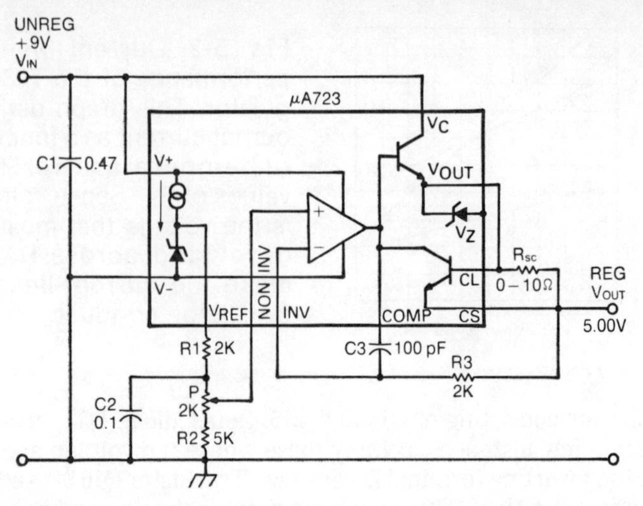

Fig. 5-3. This basic regulator circuit can be adjusted for an exact 5V output. By modifying R1, R2, and P, it can be made to produce any voltage up to 7.15V.

applications where low noise is important. Capacitor C3 is a frequency-compensating device that keeps the regulator amplifier from going into oscillation.

This regulator is capable of providing an output voltage steady to within ±0.5 mV in the face of line voltage changes up to 3V in either direction and steady to within 1.5 mV in the face of load variations up to 50 mA.

Basic Regulator with Output Greater than 7.15V

Suppose we want a regulated voltage higher than 7.15V. What do we do then? Figure 5-4 supplies an answer. The circuit resembles the one we have just considered, except that the voltage divider and R3 have been interchanged. The potential applied to the noninverting input of the error amplifier is 7.15V taken directly from the reference source, and the potential applied to the inverting input is derived from a voltage divider placed between the output and ground. With this arrangement, the minimum regulated voltage available is 7.15V (R1 = 0), and the maximum is something on the order of 3V less than the supply voltage. Of course, this doesn't mean that we can connect a 40V supply and get a continuously variable output from 7.15 to 37V. Remember that the allowable input–output differential of the μA723 is 3V. (This isn't true for all 723s. National Semiconductor's LM723 allows a 40V input–output differential.)

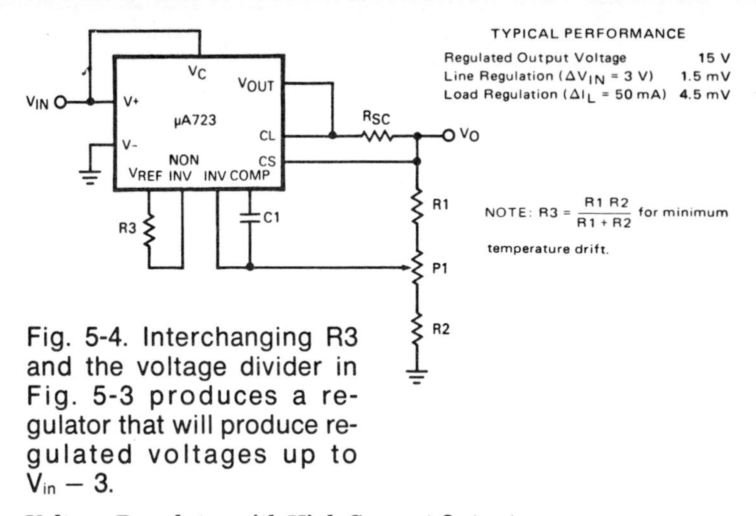

Fig. 5-4. Interchanging R3 and the voltage divider in Fig. 5-3 produces a regulator that will produce regulated voltages up to $V_{in} - 3$.

Voltage Regulator with High Current Output

Maximum output current from the 723 is 150 mA with R_{sc} equal to zero. For loads requiring greater current capacity, an output transistor with current handling ability can be used, as in Fig. 5-5A and B. For applications requiring up to 1A, Q1 could be a 2N4314 (*pnp*) or a 40347 (*npn*). The transistors used in this application should be selected on the basis of their collector current capacity and their current gain (h_{FE}). Current gain is important, since the output of the 723 regulator is limited. For a transistor with a gain of, say, 5, the most load

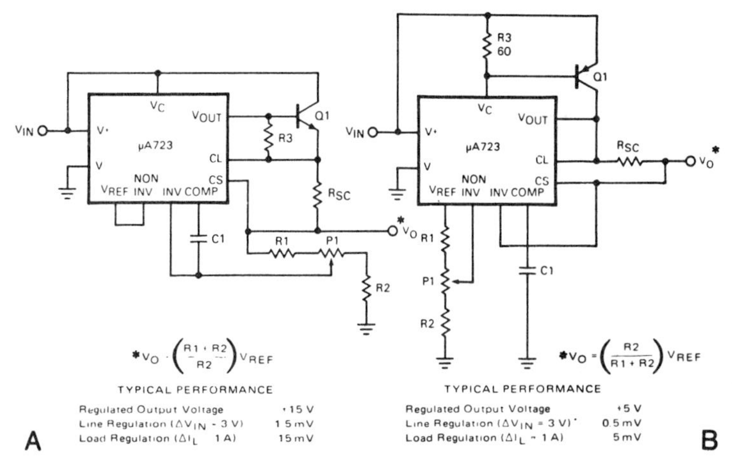

Fig. 5-5. Adding a series pass transistor to the output of the basic regulator is one way to boost available output current.

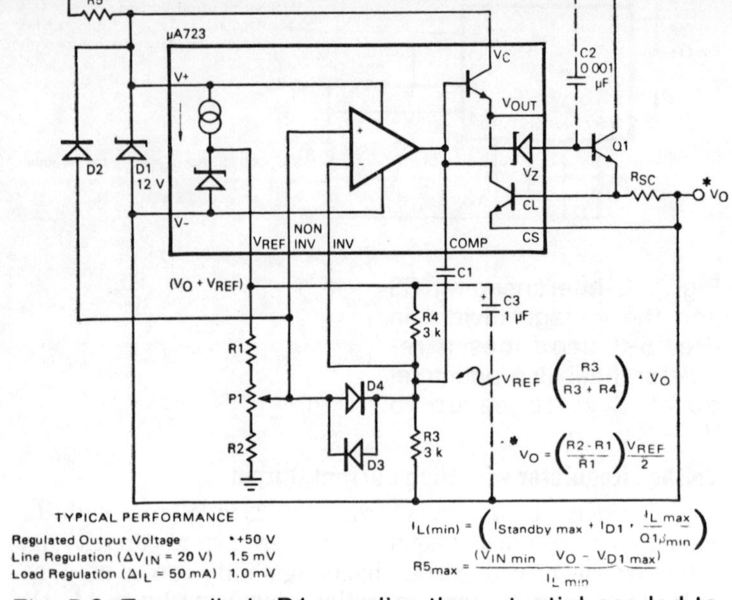

Fig. 5-6. Zener diode D1 supplies the potential needed to operate this regulator in which V_{in} is greater than 40V.

current that could be drawn without overloading the 723 would be 5 times 150, or 750 mA, regardless of the current capacity of the transistor.

Regulator with High Voltage and Current Capacity

A versatile regulator circuit capable of regulating hundreds of volts and up to several amperes of current (subject to output transistor limitations) is shown in Fig. 5-6. A 12V zener diode (D1) provides a floating power source for the IC. Diodes D2, D3, and D4 protect the regulator from excess voltages. These must be fast-acting devices. Capacitor C2 may be necessary to reduce noise level if Q1 has a wide bandwidth (high f_t). Capacitor C3 is necessary if the supply voltage is to be switched on and off after the power supply filter. If the supply voltage is switched normally, C3 will not be necessary. The figure indicates that the output voltage can be calculated from

$$V_o = \left(\frac{R2 - R1}{R1}\right)\frac{V_{ref}}{2}$$

This is only true when R3 and R4 are equal. The complete expression for V_o is

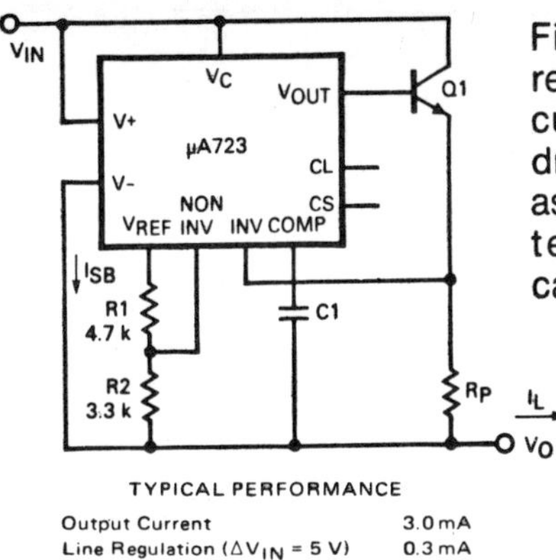

TYPICAL PERFORMANCE

Output Current	3.0 mA
Line Regulation (ΔV_{IN} = 5 V)	0.3 mA
Load Regulation (ΔR_L = 200 Ω)	0.2 mA

Fig. 5-7. In this circuit, the regulator adjusts the output current so that the voltage drop across R_p is the same as the drop across R2. The text gives formulas for calculating resistances.

$$V_o = V_{ref}\left[\frac{R2}{R1}\ \frac{R3}{R1}\ \frac{R1 + R2}{R3 + R4}\right]$$

Current Regulator

There may be applications in which a constant-current source such as the one in Fig. 5-7 may be desirable. This circuit is useful for output currents in excess of 10 mA; below that value, line regulation becomes less effective, due to the summing of the regulated current with the IC's standby current (typically, 1.3 mA). For output currents greater than 10 mA, the desired value for R_p can be calculated to a fair degree of accuracy using

$$R_p \approx \frac{3000}{I_{load} - I_{standby}}$$

where resistance is in ohms, current is in milliamperes, and where R1 and R2 have the values shown in the figure. The exact expression for R_p is

$$R_p = \frac{R2}{\left[\frac{R1 + R2}{V_{ref}}\right](I_{load} - I_{standby}) - 1}$$

Regulation to 0.005%

The precision voltage regulator of Fig. 5-8 is capable of 0.005% regulation of voltages up to the breakdown potential of the output transistors. It uses a floating 20V supply to power the IC. Output voltage is variable from zero to the output device limit. For the component values shown, output voltage V_o is

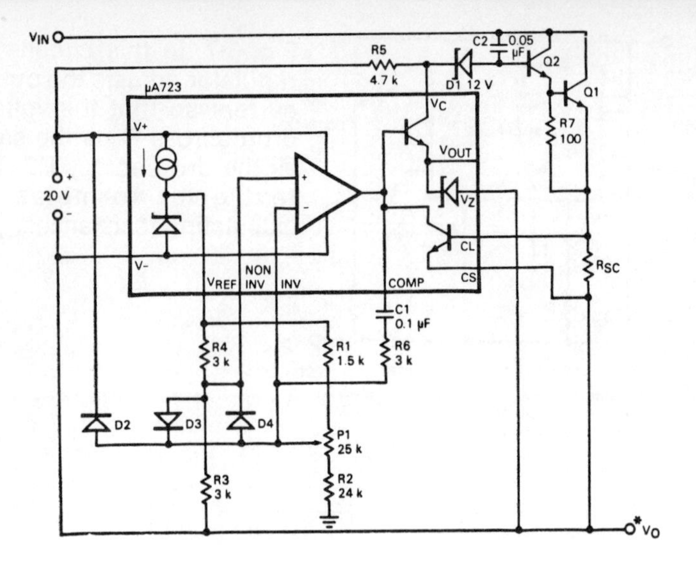

Fig. 5-8. Power supplies that use this high precision regulator circuit generally derive the 20V required to operate the 723 from a half-wave rectified auxiliary tap on their power transformer.

$$V_o = \frac{R2 - R1}{R1} \times \frac{V_{ref}}{2}$$

Foldback Limiting Regulator

Foldback current limiting gives more protection against short circuits than standard techniques. As explained in Chapter 1, the output current in a foldback limited regulator increases with decreasing load resistance only up to a predetermined point. Past that point, output current decreases rapidly, and with the output short-circuited, may be only 0.5 mA or so. Figure 5-9 shows a simple method of obtaining foldback current limiting. Selection of R1 and R2 is based on V_{sense}, the voltage necessary to forward bias the current limiting transistor. To calculate the values of R1, R2, and R_{sc} needed for any value of maximum load current (I_m) desired, use

$$R1 = V_{in} - V_{sense}$$
$$R2 = V_{sense}$$
$$R_{sc} = \frac{V_o}{I_m}\left(\frac{R2}{R1}\right)$$

where resistance values are in kilohms.

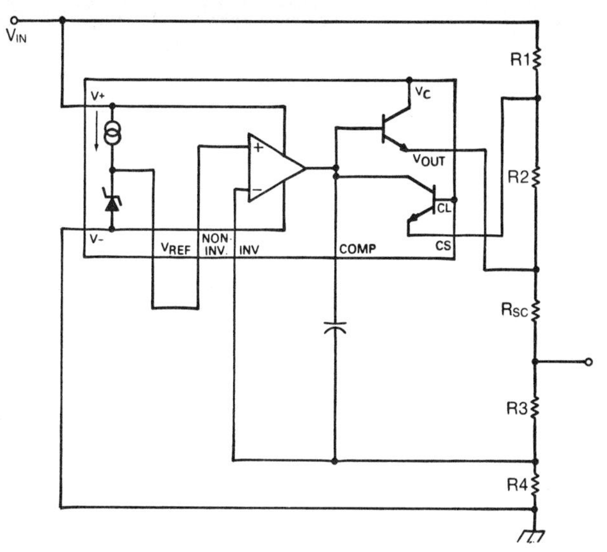

Fig. 5-9. Shorting the load on this foldback limited regulator causes the output current to go virtually to zero. When the short is removed, power must be turned off and on in order to recycle the circuit.

According to Fig. 5-2, at 25°C (room temperature), V_{sense} is 0.68 V, so R2 would be 680 ohms.

"Crowbar" Load Protection

Some voltage regulator applications may be so sensitive to overvoltages that a fast-acting sensor is required to remove all voltage from the line immediately in the event of an overvoltage. The circuit in Fig. 5-10 uses two 723 regulators, one in the ordinary regulator mode, and one as a *crowbar* sensor.

As the figure indicates, the output of the regulator is fed back to the sensor by means of potentiometer P1. This potentiometer is adjusted so that when the regulator output voltage is normal, the voltage at point *A* in the diagram is more negative than the reference voltage applied to the error amplifier inverting input. The amount by which this voltage difference is made negative is chosen so that when the regulator output exceeds a safe level, the voltage across R2 changes polarity. That is, when a safe output voltage is exceeded, the differential-mode voltage to the comparator is positive. When this occurs, two things happen almost at once, both as results of the comparator latching on. First, the SCR

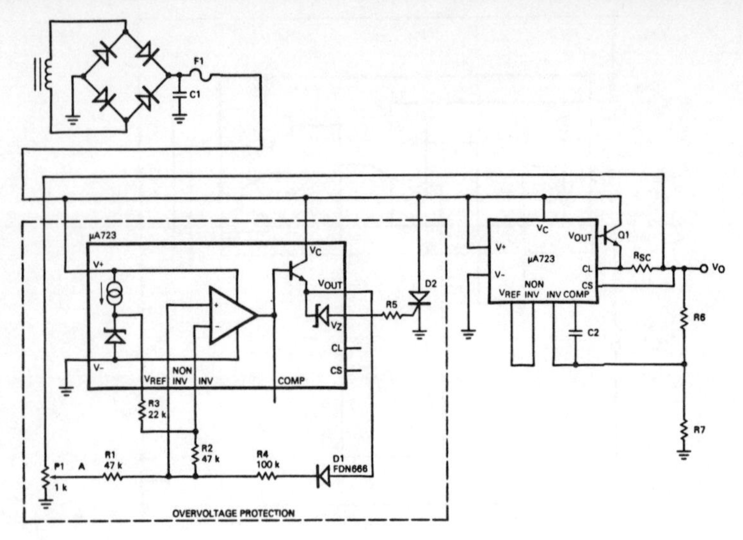

Fig. 5-10. When output voltage exceeds a preset level, the "crowbar" sensor fires the SCR, bringing the output immediately to zero and blowing the fuse. Such a drastic action might be necessary to protect computer circuits.

fires, triggered by the comparator through the zener diode. Second, the current drawn by the SCR blows the power supply fuse. This sequence of events prevents any positive voltage spikes such as might be generated if the voltage supply line were merely opened. Typically, from initial overvoltage to SCR clamping takes but 1μsec.

THE LM340

The National Semiconductor LM340 belongs to a simpler class of voltage regulator than the 723 type. As Fig. 5-11 shows, there are but three output terminals: input, output, and ground. All other connections are made right on the IC chip. A temperature-stabilized reference voltage is developed by the action of three semiconductor devices: zener D1, lateral *pnp* Q11, and *npn* Q12. The difference amplifier consists of Q13 and Q14, and the output stage is a double emitter follower consisting of Q15 and Q16.

The output voltage provided by an LM340 is designated by a suffix on the part number. For example, the output voltage of an LM340-15 is 15V. The regulator is available in both a TO-220 plastic package and a TO-3 metal case. Both are power transistor type packages, and both versions of the LM340 have an output current capability of better than 1A. If safe output

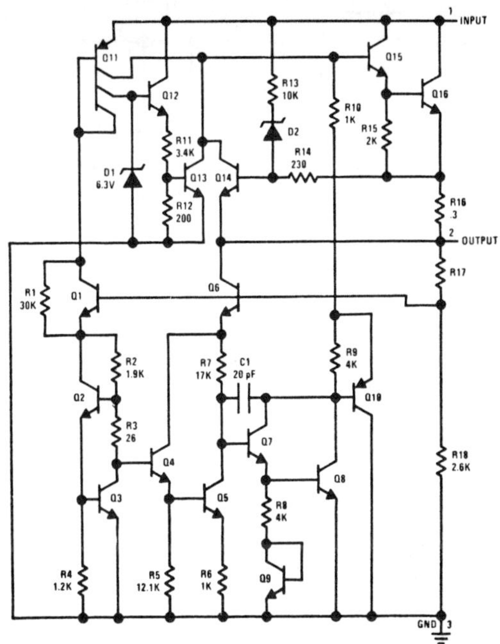

Fig. 5-11. Only three terminals are available in this in the LM340 regulator—input, output, and ground.

current level is exceeded, or if for any other reason the safe power dissipation rating of the device is exceeded, a thermal shutdown circuit takes over to protect the IC.

The LM340 is available with output potentials of 5, 6, 8, 12, 15, 18, and 24V. Allowable input voltage on all except the -24 model is 35V; on the LM340-24, it is 40V.

Using the LM340 is very easy, as the dual-voltage supply in Fig. 5-12 shows. The diodes assure startup into a common load regardless of input voltage startup sequence.

THE LM109

The LM109 is a three-terminal 5V positive regulator similar in function and design to the LM340. In its TO-5 package, it can deliver currents up to 200 mA. In its TO-3 version, available output current is greater than 1A.

Improved Positive and Negative Regulators

Typical LM109 voltage regulation error is 1%. This can be improved, using the circuit of Fig. 5-13, to a regulation better than 0.01%. An LM108 op-amp compares the output voltage to a zener reference and derives an error signal that is used to

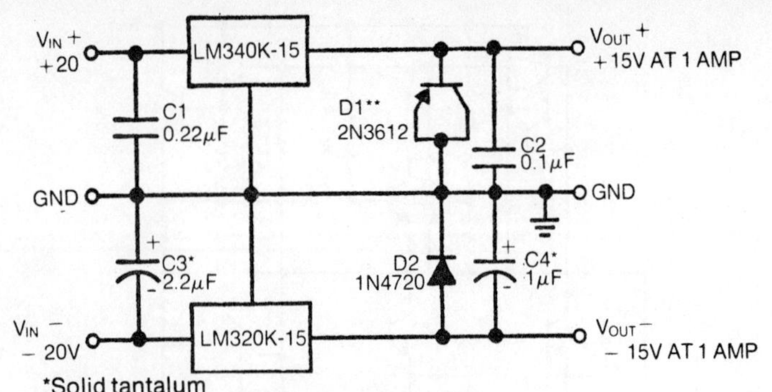

Fig. 5-12. Using the LM340 for positive or negative voltages is simplicity itself.

drive a *p*-channel junction FET placed in series with the LM109 ground lead.

A negative voltage regulator using a similar LM108 error amplifier is shown in Fig. 5-14. This version uses discrete components rather than the LM109 of the positive regulator. Field-effect transistor Q1 level-shifts the output of the error amplifier to drive output transistors Q3 and Q4. Transistor Q2 provides current limiting.

Some construction precautions relative to both of these regulators are worth noting. For all the polarized capacitors in the drawings, solid tantalum units must be used. Unlike

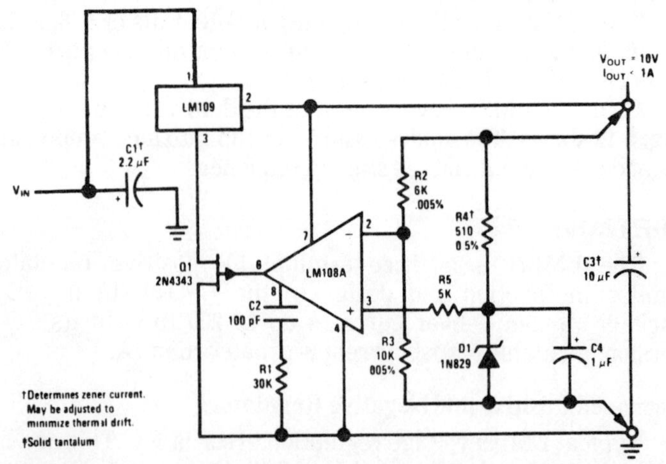

Fig. 5-13. The regulation of the LM109 can be extended using an LM108 op-amp. The LM108's extremely low input current is critical to proper operation of this circuit.

ordinary electrolytics, tantalum capacitors have a satisfactory low impedance at high frequencies. The resistors must be low-temperature-coefficient wirewound or precision metal-film units. Even 1% carbon, metal-film, or tin-oxide resistors are unsuitable because of their temperature drift. Cheaper zener diodes should not be substituted for the 1N829 device specified, or again, thermal drift will interfere with regulation. Finally, note that both drawings specify sensing *at the load*. National Semiconductor warns that even 1 in. of wire between load and sensing input will degrade regulation. All ground leads should be returned to a single point, as shown, to preclude ground loops. Shielding of the zener, error amplifier, and voltage sensing resistors is also recommended.

THE CA3085

A typical RCA voltage regulator, the CA3085, is shown in Fig. 5-15. Supplied in an 8-pin TO-5 package, the regulator may be had in any of three versions, the basic CA3085, the CA3085A, and the CA3085B. The basic version has an input range of 7.5 to 30V, an output range of 1.8 to 36V, and a maximum output current of 100 mA. Input range of the CA3085A and B each have a maximum output current of 100 mA. Input range of the CA3085A is 7.5 to 40V, with an output range of 1.7 to 36V; the

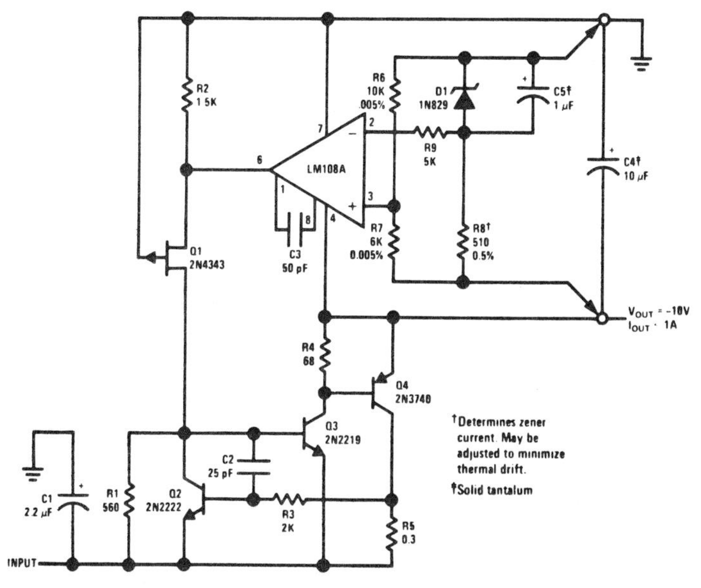

Fig. 5-14. A negative regulator, also using the LM108. The text points out a number of precautions to be taken when constructing these circuits.

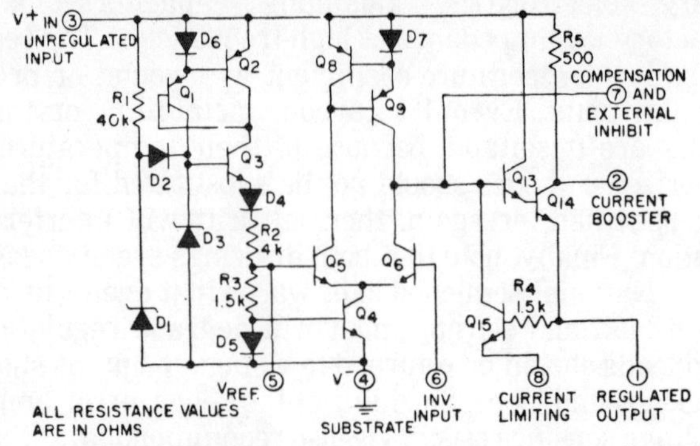

Fig. 5-15. RCA's CA3085 has a lower reference voltage than the 723, extending its usefulness.

CA3085B extends this range by 10V. Maximum safe input differentials for the basic version, the model A, and the model B are 30, 40, and 50V, respectively.

In general, the circuit of the CA3085 has much in common with the 723 regulator. Note, however, that the reference voltage is applied to the noninverting input of the error amplifier within the IC, making it necessary that the feedback voltage divider always be connected to the regulator output terminal.

CA3085 Regulated Power Supply

As Fig. 5-16 shows, the CA3085 is similar in application to the 723. The low (1.6V) reference potential permits great latitude in output voltage. The 5.6Ω resistor in the figure is R_{sc}, the short-circuit protection resistor.

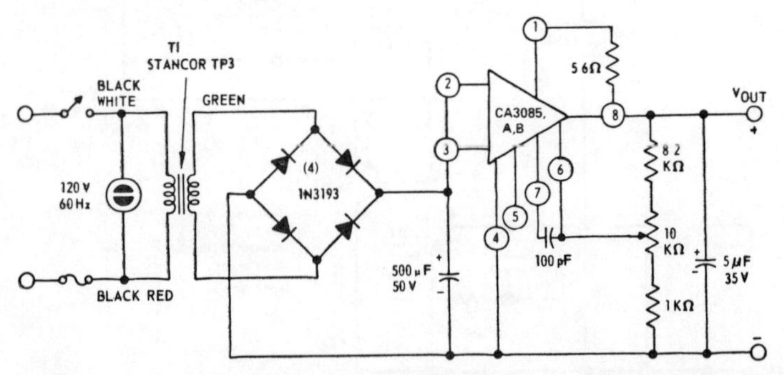

Fig. 5-16. The CA3085 is used in much the same manner as the 723. The more complicated circuits using the 723 that we covered in the first part of this chapter could also have used a CA3085.

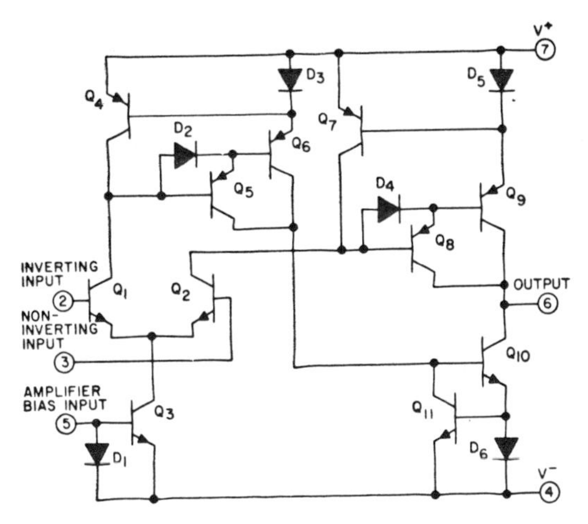

Fig. 5-17. The CA3080 is actually just the front two-thirds of a CA3094 (compare to Fig. 2-23).

THE CA3080

RCA's CA3080 (Fig. 5-17) is an operational amplifier, the circuit of which resembles the first two-thirds of a CA3094 (Fig. 2-23). Figure 5-18 pictures the output stage that the CA3094 adds to the CA3080. Both devices constitute what RCA calls *operational transconductance amplifiers*. That is, the gain of these devices is a linear function of the bias current applied to pin *3*.

900V Supply

The high voltage supply in Fig. 5-19 uses both a CA3094 and a CA3080. The output voltage is high, about 900V, but current capacity is low, about 26 μA being maximum. This sort of supply is typically used to generate the high voltage needed by

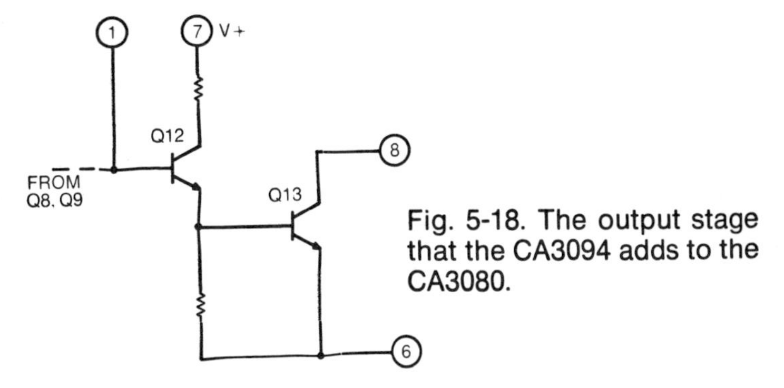

Fig. 5-18. The output stage that the CA3094 adds to the CA3080.

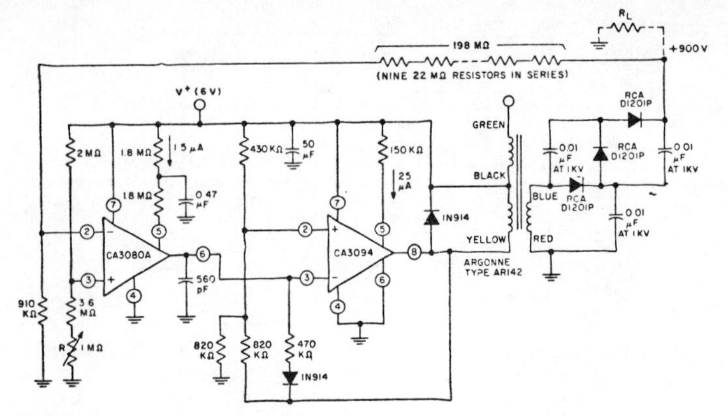

Fig. 5-19. This high-voltage supply is unlike any of the regulators considered so far in this chapter in that it uses feedback to control the frequency of an AC signal that is stepped up in voltage by the transformer.

the Geiger-Mueller tube in a radiation detector. Output voltage regulation is about 1% from 5 to 26 μA output current. DC-to-DC conversion efficiency is good—about 48%.

The two op-amps form a square-wave oscillator, the frequency of oscillation being based on the charging rate of the 560 pF capacitor. This is set by the output current from the CA3080, which is itself a function of the output voltage fed back through the lopsided voltage divider and a reference voltage set by the 1M pot. The transformer is a cheap audio unit, half its 200Ω secondary winding connected between the 6V power line and the CA3094 output transistor collector. Output is taken from the transformer's 500K primary. The approximately 300V of transformer output is stepped up and rectified in a conventional voltage tripler.

Regulating High Voltage with a Low-Voltage Transistor

The regulator in Fig. 5-20 uses a relatively low-voltage transistor to control a high output voltage. This is possible because 200V zener diode D1 absorbs most of the supply voltage. Note that unlike other regulators in this chapter, which are series regulators, this is a shunt regulator—one in which current is dissipated in a circuit parallel to the load so as to keep the voltage constant at one end of a resistor (R1).

Generally, shunt regulators are not used because their greater current drain makes them less efficient than series regulators. In this case, however, the shunt regulator is capable of handling more voltage with a smaller transistor

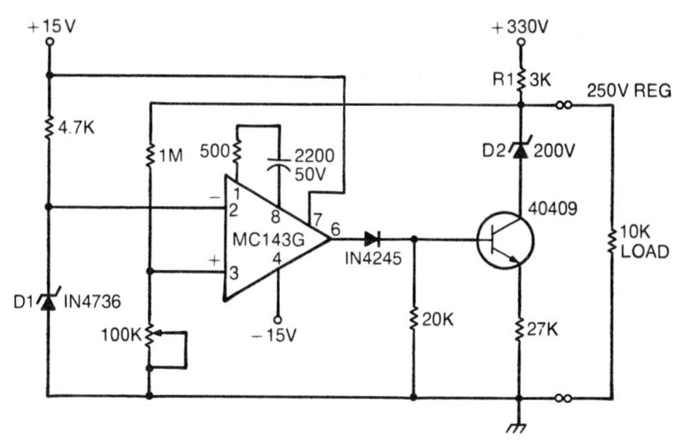

Fig. 5-20. Shunt regulation of a 330V supply is here accomplished with a 40409, whose breakdown voltage is 90V, through the use of a 200V zener in series the 40409.

than an equivalent series regulator, overcoming its disadvantage.

The actual reference supply in the circuit is simply a zener diode, making this regulator less effective than others we have seen. The principle involved in the output stage, of course, is applicable to more sophisticated regulators as well.

Zero-Ripple Regulator

The simple series regulator in Fig. 5-21 adds an interesting feature to an otherwise conventional regulator circuit. The result of this addition is a power supply output with virtually

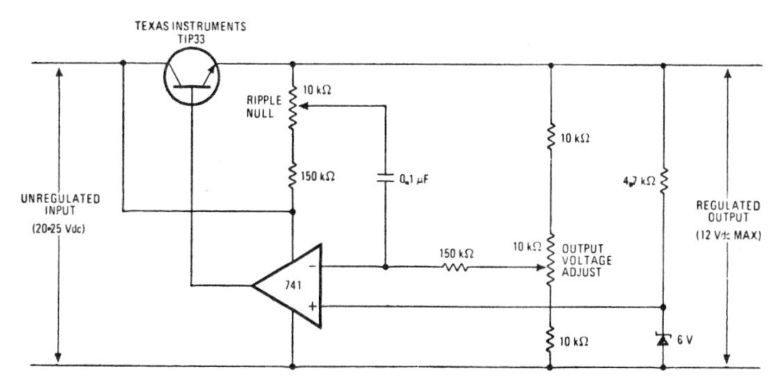

Fig. 5-21. By feeding back a signal 180° out of phase with line ripple, this circuit produces a ripple-free regulated output.

no ripple. A portion of the unregulated ripple-carrying input voltage is fed to the error amplifier's inverting input via the *ripple null* pot and the 0.1 μF capacitor. This signal is amplified and applied to the *series pass* transistor 180° out of phase with the input ripple. The result is a ripple-free output.

In operation, the output voltage is set to the required level and a "worst-case" load is applied. Then, using an oscilloscope to monitor the output voltage, the ripple null pot is adjusted for minimum ripple. The ripple will remain essentially zero for any load less than the maximum.

Measuring Equipment

The linear IC operational amplifier, with its high input impedance and good linearity can be used in voltage, current, and even, as we shall see, resistance-measurement circuits. Another linear IC, the analog multiplier, can be used in a wattmeter circuit.

THE LM112

Like the LM108, National Semiconductor's LM112, (Fig. 6-1) uses superhigh-gain IC transistors to achieve very low input-offset voltage and input-current errors. The circuits of the two op-amps are similar (the LM108 appears in Fig. 3-22), but the LM112 features internal compensation and provision for offset balancing. Feed-forward compensation is possible using a 1000 pF capacitor between pin 5 and ground. The LM112 requires dual supply voltages between ±2 and ±20V. Package options are 8-pin TO-5, 10-pin flat-pack, and 14-pin DIP.

Temperature Sensor

The temperature sensor in Fig. 6-2 can be calibrated for an output scale factor of 100 mV/°C with better than 1% error over a 100° range. Its design is based on the emitter–base turn-on voltage variation of an inexpensive silicon transistor. In silicon transistors, this turn-on voltage varies linearly with temperature.

Two LM113 high-precision zener diodes regulate the input voltage to the LM112 op-amp, which is connected in its inverting mode. The extremely low input bias current of the LM112 contributes substantially to overall accuracy.

In operation R4 is adjusted to zero the amplifier output when the 2N2222 is at 0°C (the temperature of melting ice). Then R5 is adjusted so that the amplifier output is 10V when the 2N2222 probe is at 100°C (the temperature of boiling water at standard atmospheric pressure—29.92 in. of mercury). If you are using a barometer to determine atmospheric pressure, remember that the barometer reading must be corrected for

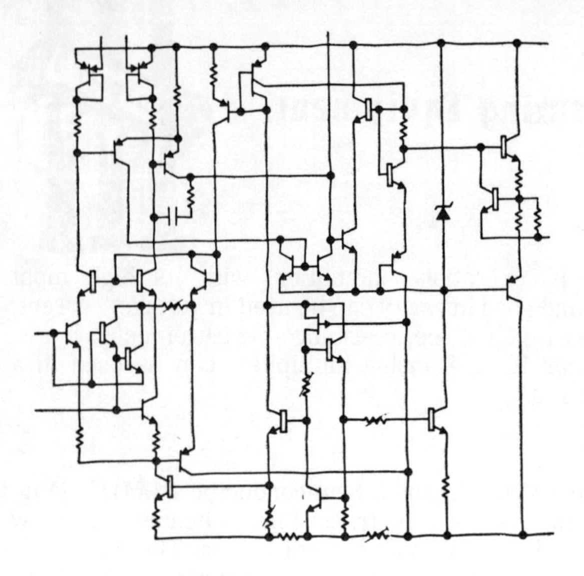

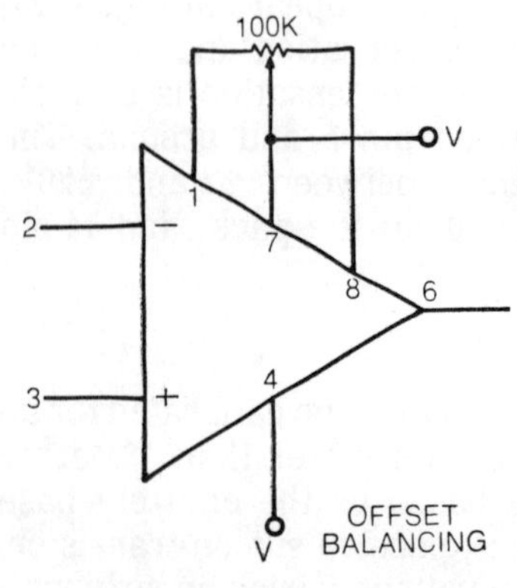

Fig. 6-1. The LM112 uses standard IC transistors (fat bases) and superhigh-gain transistors (thin bases) to achieve high gain and low input bias and offset current. It is fully compensated and can be offset nulled with a 100K potentiometer as shown.

your altitude. If you live at an elevation 1000 ft above sea level, a calibrated barometer will indicate the pressure that would exist in your locale at the bottom of a 1000 ft hole.

THE LM4250

The LM4250 (Fig. 6-3) is an op-amp with several interesting features. For one thing, it resembles RCA's operational transconductance amplifiers in that its gain may be controlled by means of an external bias current (into pin 8). National Semiconductor points out that this input also controls such second-order effects as input bias offset, quiescent power consumption, slew rate, input noise, and

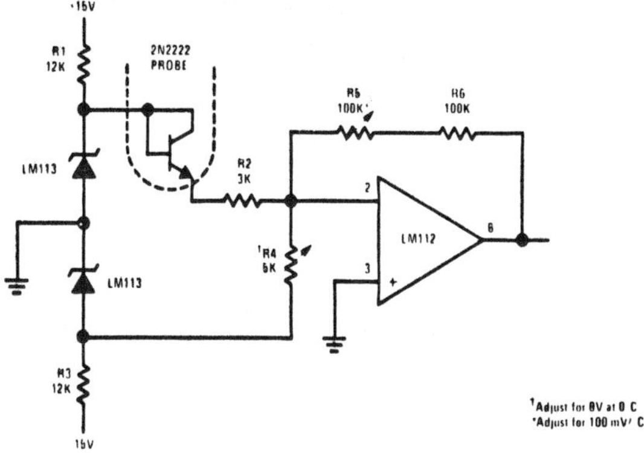

Fig. 6-2. The base–emitter turn-on voltage of a silicon transistor varies linearly with temperature. This behavior and a highly stable 1.2V zener reference lead to a temperature sensor with 1% accuracy.

gain–bandwidth product. The LM4250's second remarkable characteristic is its extremely low supply voltage and negligible quiescent power requirement. With a supply voltage range from ±1 to ±18V and a quiescent current drain of 7.5 μA under certain bias conditions, an LM4250 can operate for

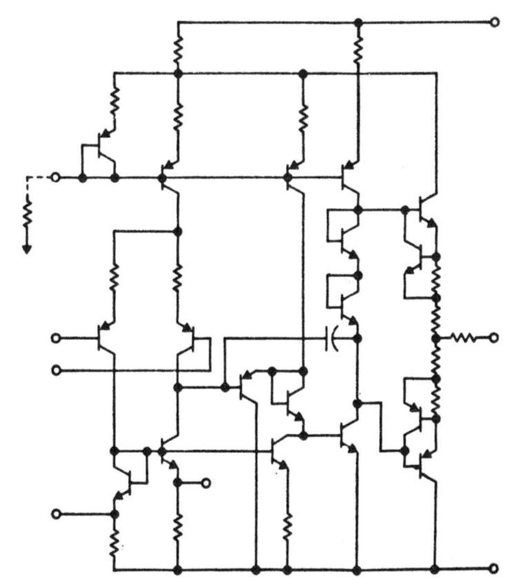

Fig. 6-3. Operating from supply potentials as low as ±1V, the LM4250 nonetheless provides all the performance one would expect from a quality op-amp. Moreover, its gain can be controlled by a current applied to pin 8. The LM4250 lists for less than five dollars.

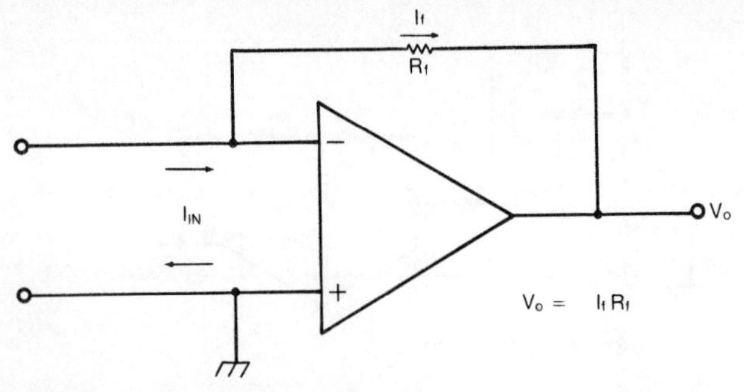

Fig. 6-4. An operational amplifier maintains an output voltage such that I_f always equals I_{in}. If we know the value of R_f, then we can calculate I_{in} from V_o. Since it's easier to measure a large V_o than a small I_{in}, this leads to a practical, cheap nanoammeter.

as long as a year from two "D" cells. Package options are an 8-pin TO-5 and an 8-pin DIP.

Current- and Voltage-Measuring Circuits

In Chapter 1, we looked at the behavior of operational amplifiers with negative feedback, such as the one in Fig. 6-4. We noted that because of the op-amp's high input impedance, the input current, I_{in}, must equal the feedback current, I_f, and that, because of the extremely high gain of the amplifier, the potential difference between the inverting and noninverting inputs with the amplifier nonsaturated must be no more than a few microvolts, a level we could conveniently regard as zero. Given these conditions, we see that for the circuit in Fig. 6-4, the output voltage is equal to the input current, I_{in}, times the feedback resistance, R_f, times -1. This gives us a means of measuring extremely small currents with relatively crude and inexpensive voltmeters. If we make R_f large enough, even a small current will produce a substantial voltage at the op-amp output.

The accuracy of such a current measuring arrangement, of course, depends upon how closely the operational amplifier matches the characteristics of an ideal operational amplifier. For this sort of application, the LM4250, with its 7.5-nanoampere (7.5 nA) input bias current and its 92 dB gain, is close to ideal.

Figure 6-5 shows an arrangement for measuring currents down to 100 nA (a nanoampere equals 10^{-9} ampere, or 0.001 μA). Diodes D1 and D2 protect the inputs against excessive voltages. Bias resistor R_s sets the open-loop gain of the

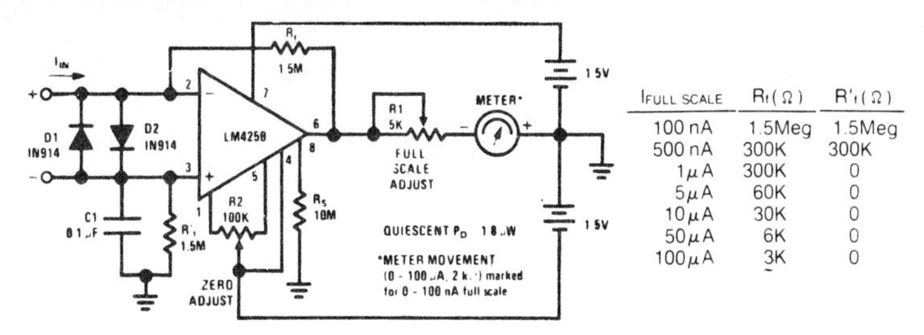

$I_{FULL\ SCALE}$	$R_f\ (\Omega)$	$R'_f\ (\Omega)$
100 nA	1.5Meg	1.5Meg
500 nA	300K	300K
1 μA	300K	0
5 μA	60K	0
10 μA	30K	0
50 μA	6K	0
100 μA	3K	0

Fig. 6-5. A practical nanommeter using an LM4250.

LM4250, potentiometer R2 zeros the meter by nulling the offset, and resistor R_f provides the feedback that sets the scale factor, as indicated by the table. An additional resistor, R_f', equal to R_f, is used for measurements below 1 μA to minimize error resulting from input bias currents. Note that with R_f' connected, output voltage is equal to $-2(R_f I_{in})$.

To measure higher currents, from 1 mA to 10A, the circuit in Fig. 6-6 uses a shunt resistor, R_A, in much the same manner that a conventional ammeter does. The shunt resistors in this circuit, however, have relatively high resistances, making them cheaper and easier to obtain than conventional shunts.

The circuit in Fig. 6-7 makes a DC voltmeter of the LM4250 by adding an input resistor, R_v. This resistor is the primary element determining input resistance, so this meter's resistance varies from 10 megohms per volt (M/V) at the lowest range to 100,000 Ω/V in the 100V range. Resistor R_f' is added in some ranges for the same reason it was added in the low-current ammeter. Diodes D1 and D2 protect against overvoltages of as much as 500V in the 100 mV range.

In all three of these op-amp meter circuits, resistor precision is critical: 1% or better film or wirewound resistors are essential.

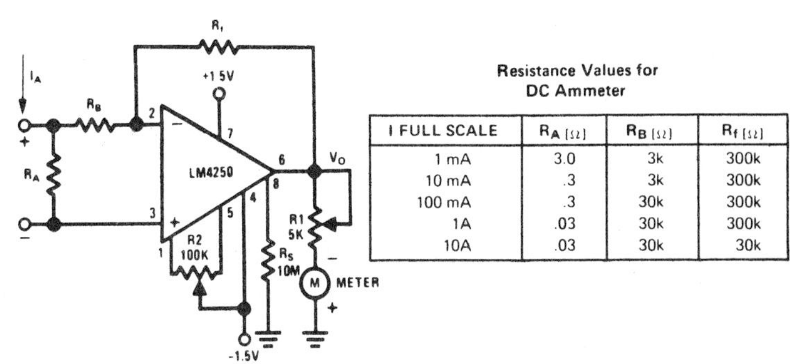

Resistance Values for DC Ammeter

I FULL SCALE	$R_A\ (\Omega)$	$R_B\ (\Omega)$	$R_f\ (\Omega)$
1 mA	3.0	3k	300k
10 mA	.3	3k	300k
100 mA	.3	30k	300k
1A	.03	30k	300k
10A	.03	30k	30k

Fig. 6-6. Splitting the input current between R_A and R_B extends the range of the nanoammeter to 10A.

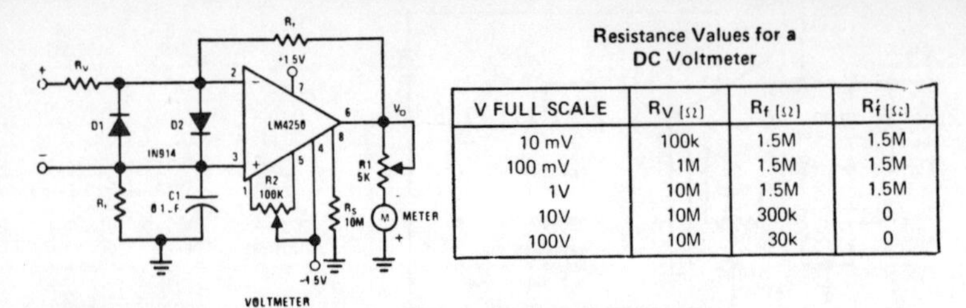

Resistance Values for a DC Voltmeter

V FULL SCALE	Rv (Ω)	Rf (Ω)	Rf′ (Ω)
10 mV	100k	1.5M	1.5M
100 mV	1M	1.5M	1.5M
1V	10M	1.5M	1.5M
10V	10M	300k	0
100V	10M	30k	0

Fig. 6-7. An input resistor turns the nanoammeter into an electronic voltmeter.

THE LH0042

For even lower input bias and offset currents than the LM4250, there is the LH0042, with its FET input stage (Fig. 6-8). Unlike the LM4250, the LH0042 requires a substantial supply voltage: ±15V typical, with 2.4 mA quiescent drain. The device is packaged in 8-pin TO-5 cans, 10-pin flat-packs, and 14-pin DIPs.

A Different Approach to Voltmeter Circuitry

Operating as a voltage- rather than a current-sensitive device, the high-resistance DC voltmeter of Fig. 6-9 superficially resembles the previous project. Note, however, that this op-amp is used in its noninverting mode and that the range switch selects taps on a voltage divider. The resistors in the divider are selected so that the full-scale input voltage always produces 100 mV at the amplifier input. The 1M

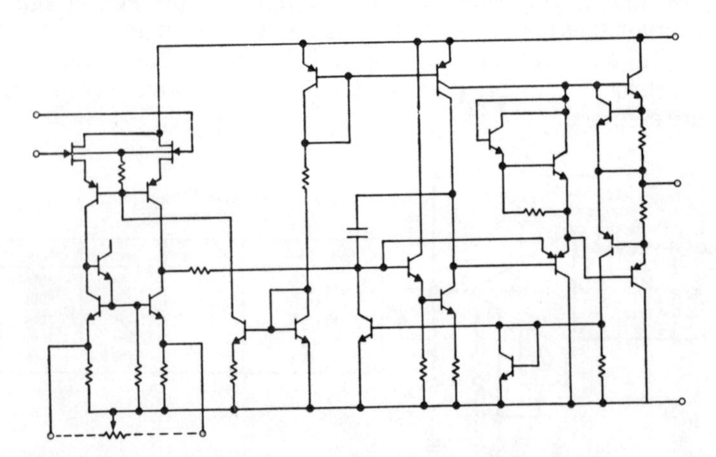

Fig. 6-8. The LH0042 derives its extremely low input bias and offset characteristics from its FET input stage.

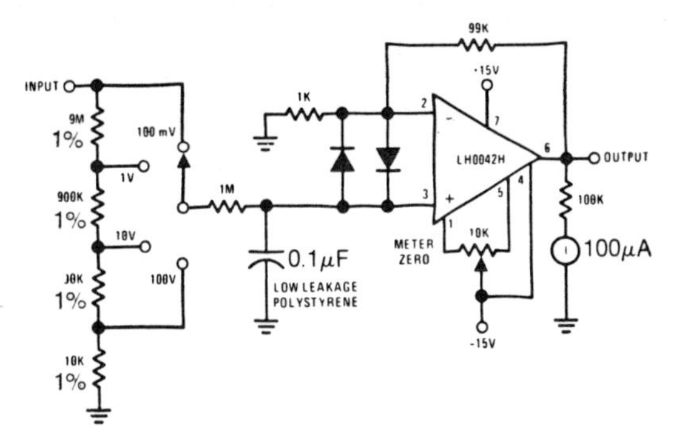

Fig. 6-9. This electronic voltmeter uses an op-amp stage with a gain of 100 and a voltage divider input.

resistor's function is to maintain a high input resistance (at least 10 M/V) to the voltmeter. The gain of the op-amp is fixed by the 99K feedback resistor and the 1K input resistor at 1 + 99, or 100. This produces a 10V output at full scale. The 100K resistor adapts this to the 100 μA meter.

An AC Voltmeter With a Novel Approach

For accurate measurement of AC voltages, the circuit of Fig. 6-10 can be used. The gain of the LH0062 op-amp is set for unity by R7 and R10. The current flowing through the bridge rectifier is equal to the op-amp's output voltage divided by the resistance selected by the range switch. This gives a full-scale

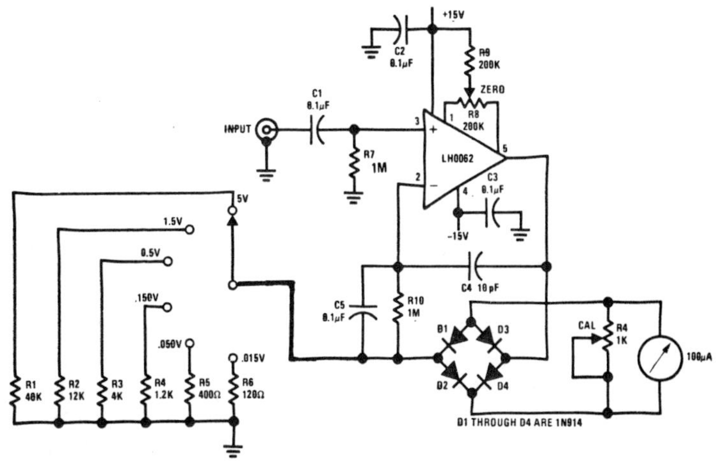

Fig. 6-10. This AC/DC voltmeter uses still a different approach—unity gain and a choice of resistors between output and ground to scale the microammeter current.

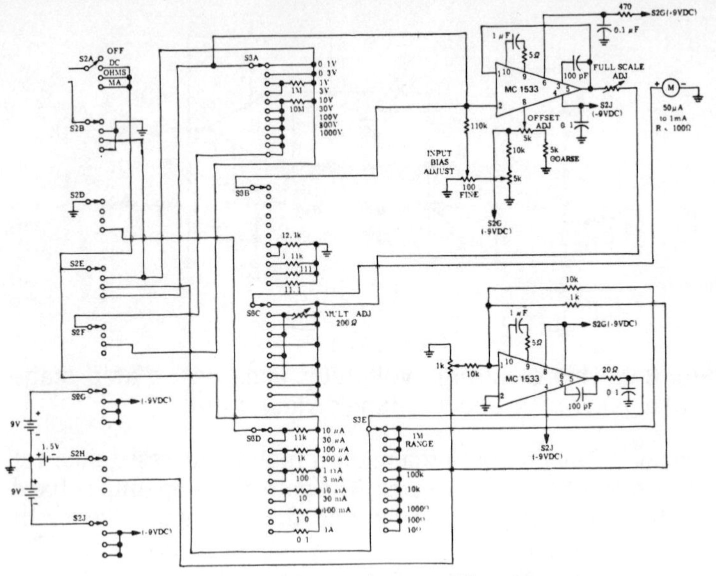

Fig. 6-11. Complete multimeter.

current of 125 μA in each range position. Calibration shunt resistor R4 is adjusted to reduce this to 100 μA through the meter. The op-amp provides a buffer between the voltage being measured and the meter, the input resistance of the voltmeter op-amp being at least 200,000 Ω/V.

A Complete VOM

Motorola suggests a circuit (Fig. 6-11) that constitutes a complete linear IC VOM. The buffer/meter driver is an MC1533 op-amp used as voltage follower. In the voltmeter mode, it responds to voltages developed across a voltage divider formed by the resistors in banks A and B of S3. The *multiplication adjust* pot on bank C of S3 allows meter ranges between decade increments. In the ammeter mode, voltage drops across the shunt resistors of S3D provide the input to the voltage follower.

The ohmmeter function requires an additional MC1533 op-amp, connected as in Fig. 6-12. A zero-adjust voltage is applied to the inverting input of this op amp with the ammeter probes shorted together, and the pot is adjusted for a full-scale meter reading. Then the unknown resistor is inserted between the ohmmeter probes, and the voltage at the output of a voltage divider (formed by the unknown and one of the S3D resistors) is applied to the voltage follower.

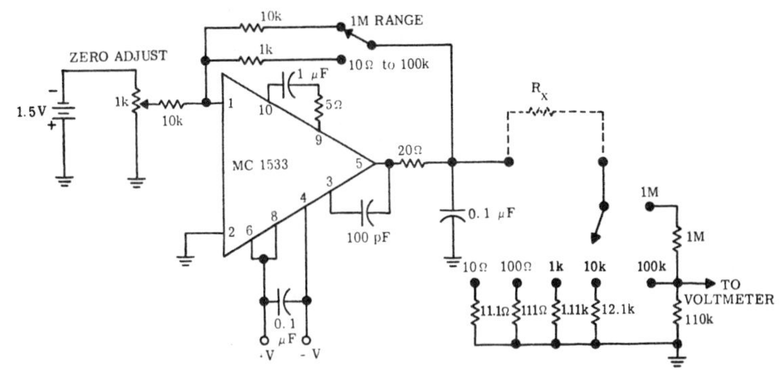

Fig. 6-12. The ohmmeter is zereoed with the resistance probes shorted (zero ohms is full-scale meter deflection), then meter reading corresponds to potential at output of voltage divider.

Wideband Wattmeter

A wattmeter with frequency response from DC to several kilohertz and measurement capability up to 2 kVA can be made using a Signetics 5596 analog multiplier, as in Fig. 6-13. Op-amp A1 produces a voltage output proportional to the current flowing through the load. This is based on the voltage drop across shunt resistor R1. The current analog signal is applied to the V_x input of the multiplier.

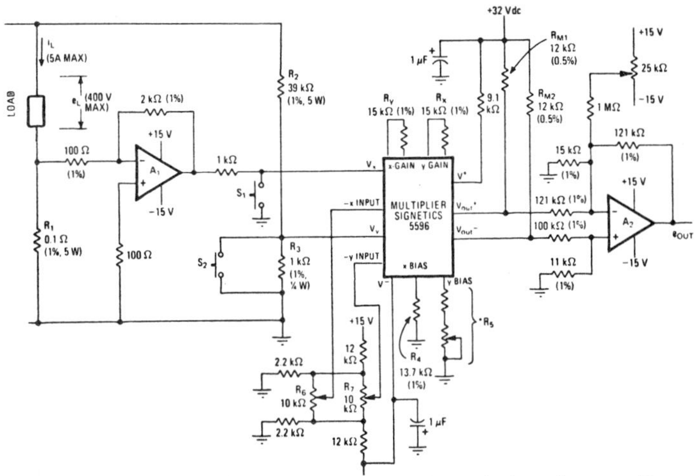

Fig. 6-13. Op-amp A1 generates a signal that is a voltage analog of the current through the load. This is multiplied by a voltage signal representing the load voltage to produce a signal representing power dissipated in the load.

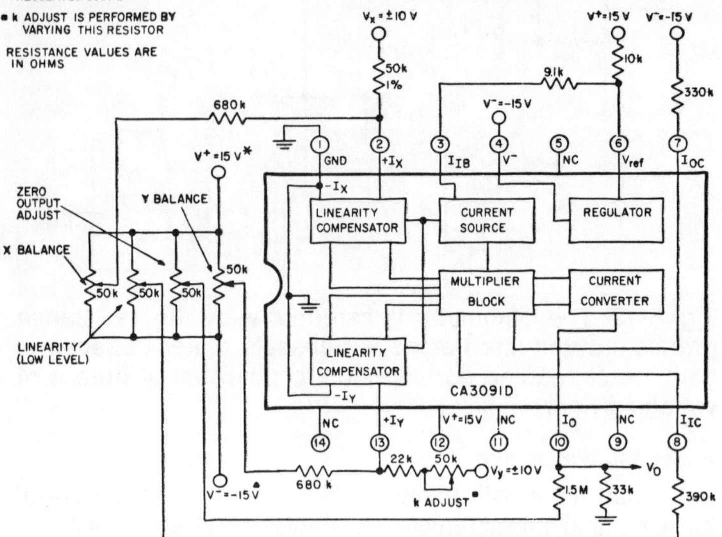

Fig. 6-14. RCA's CA3091D is another analog multiplier that can be used in the wattmeter circuit of Fig. 6-13. It is shown here with its required external circuitry. Note that the actual IC multiplicand inputs are current ports.

A load-voltage signal is derived from voltage divider R2–R3 and applied to the multiplier's V_y input. The output of the multiplier is a double-ended signal proportional to V_x times V_y. This signal is buffered, level-shifted (in the unity-gain output amplifier A2), and read with a voltmeter. With the component values in the figure, the output voltage factor is 5 mV/VA (millivolts per voltampere). This factor can be increased by changing R1 and R2–R3 if this factor produces inconvenient readings in your applications. Note that since the output signal will have the same frequency as the line voltage to the load, you will have to use a voltmeter with a suitable bandwidth to read the wattmeter output.

Another analog multiplier that can be used for this purpose is RCA's CA3091D, shown in Fig. 6-14 with the external resistors it requires.

Filters

Active filters using linear integrated circuits have several advantages over conventional RLC filters. For one thing, the filter designer can use the extremely high gain of the operational amplifier to boost the effective Q of his circuit or to provide more than unity (0 dB) gain. For another, he may use voltage followers to provide isolation between input and output of a conventional filter, simplifying his design by isolating the filter elements from a low-impedance or reactive load. Finally, by proper use of op-amp feedback, he can eliminate coils from his filters, reducing them in size, weight, and cost.

ONE-POLE LOW-PASS AMPLIFIER

Figure 7-1A shows a generalized low-pass amplifier using a single op-amp, three resistors, and a capacitor. The graph in Fig. 7-1B shows its performance. The response of the amplifier was determined from the basic gain expression for the operational amplifier:

$$\frac{V_o}{V_i} = \frac{Z_{fb}}{Z_i}$$

where

V_o = output voltage
V_i = input voltage
Z_{fb} = feedback impedance
Z_i = input impedance

In this case, the feedback impedance is the impedance of the parallel combination of R3 and C1, or

$$Z_{fb} = \frac{R3}{j\omega\,(R3C1 + 1)}$$

and the input impedance is just R1. Thus, the gain of the circuit is

$$\frac{R3}{R1}\left[\frac{\frac{1}{R3C1}}{j\omega + \frac{1}{R3C1}}\right]$$

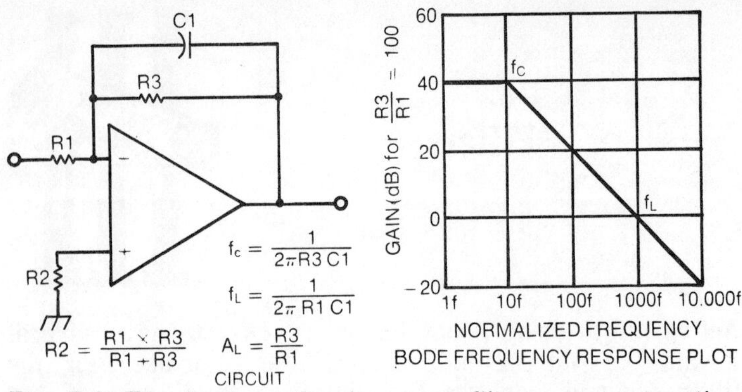

Fig. 7-1. The basic active low-pass filter uses a reactive feedback element and a resistive input element.

The 3 dB break point, f_c, corresponds to the radian frequency (ω) at which the magnitude of the gain expression is half its maximum value. From this frequency, the gain rolls off at 20 dB/decade. The unity gain frequency, f_L, is the frequency at which the amplifier gain reaches 0 dB. These two points are plotted in the figure for the case where the ratio of R3 to R1 is 100. Changing this ratio would change the spacing between f_c and f_L, but would not change the 20 dB/decade rolloff.

TWO-POLE LOW-PASS FILTER

A different kind of low-pass filter with a maximum gain of 0 dB is shown in Fig. 7-2A. This filter features a National Semiconductor LM102 voltage follower, an op-amp with 100% feedback, and all required compensation built right into the chip. The LM110, an improved version of the LM102, could also be used in this filter. If the advantage of using a unity gain amplifier in a filter isn't immediately apparent, remember that the voltage follower is primarily an impedance-changing device. The performance of a simple RC low-pass filter is severely compromised if it is connected to a low-impedance or reactive load. The circuit in the figure has the characteristics of two isolated filter sections; that is, it has a 40 dB/decade rolloff. Its output impedance is 0.8Ω, low enough to drive any load.

Filter design starts with choosing the cutoff frequency, f_c, and arbitrary values for R1 and R2. These should be somewhere in the range from 10K to 10M. Then C1 and C2 can be calculated using

$$C1 = \frac{R1 + R2}{(8.88) f_c \, R1R2}$$

$$C2 = \frac{0.225}{(R1 + R2)\, f_c}$$

If this yields unacceptable values for C1 and C2, R1 and R2 can be changed and new capacitance values computed. Note that if R1 and R2 are equal, C1 will be exactly twice as big as C2.

LOW-PASS AMPLIFIER WITH HIGHER Q

The National Semiconductor LM3900 "Norton" amplifier in Fig. 7-2B is capable of gain, so the low-pass filter it forms a part of, while superficially similar to the previous filter, gives us something extra to play with in terms of gain and Q.

Designing a low-pass filter such as this, we begin by listing our requirements for passband gain A (the feedback resistance ratio), break frequency, f_c, and Q. Then, choosing an arbitrary value for C1, we define an arbitrary constant, k, such that

$$C2 = k\, C1$$

We might start out with k equal to 1, and if we find that the values we calculate for R2 and R3 are unreasonably large, we can redefine k as a larger value and compute new, smaller values for R2 and R3.

To compute R2, we use

$$R2 = \frac{1}{2Q\, 2\pi f_c\, C1}\left[1 \pm \sqrt{1 + \frac{4Q^2\,(A + 1)}{k}}\right]$$

Then, $R1 = R2/A$, and

$$R3 = \frac{1}{(2\pi f_c)^2\, C1^2\, R2\, k}$$

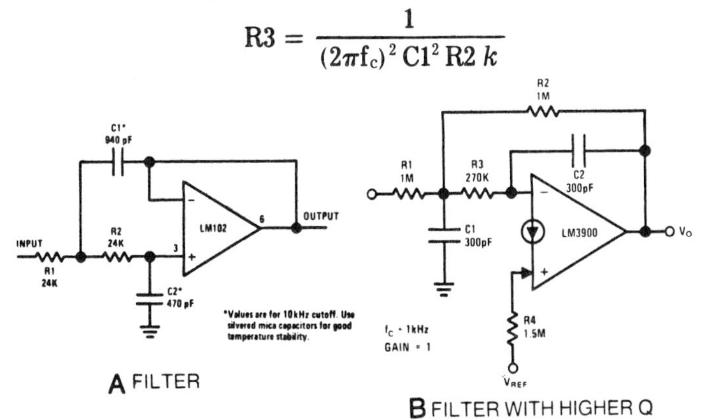

A FILTER

B FILTER WITH HIGHER Q

Fig. 7-2. Two practical low-pass filters, one using a voltage follower (A), the other using one amplifier from a low-cost "Norton" amp (B). Design equations are given in the text.

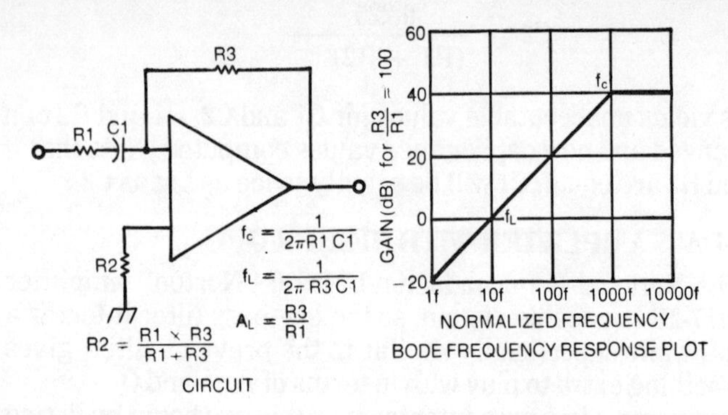

Fig. 7-3. This amplifier uses a reactive input and a resistive feedback element to achieve the response characteristic pictured.

ONE-POLE HIGH-PASS AMPLIFIER

A high-pass filter can be realized in a manner similar to the low-pass filter of Fig. 7-1. The circuit diagram for a one-pole active filter is shown in Fig. 7-3A. Its response is graphed in Fig. 7-3B. In this case, the feedback impedance is just R3, but the input impedance Z_i is the equivalent of R1 in series with C1 or

$$Z_i = R1 \frac{j\omega + \frac{1}{R1\,C1}}{j\omega}$$

Gain is the ratio of feedback impedance to input impedance, or

$$A = \frac{R3}{R1}\left[\frac{j\omega}{j\omega\ \frac{1}{R1C1}}\right]$$

As in the case of the low-pass filter, the rolloff rate is 20 dB/decade, but now the unity gain frequency, f_L, is lower than the cutoff frequency, f_c.

TWO-POLE HIGH PASS FILTER

The voltage-follower high-pass filter (Fig. 7-4A) interchanges the locations of resistors and capacitors, but its design remains the same as that of the low-pass filter in Fig. 7-2A. The design frequency of the filter in the diagram is 100 Hz. This illustrates another advantage of an active filter over a passive filter. For any normal value of source and load

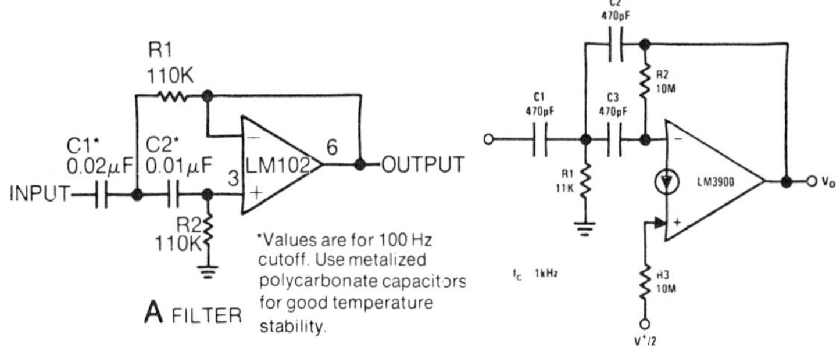

Fig. 7-4. These high-pass filters interchange the positions of resistors and capacitors in the low-pass filters of Fig. 7-2. For the voltage-follower version, the design equations are unchanged. For the higher-Q version, the design equations are somewhat different.

impedance, a 100 Hz cutoff frequency would just not be obtainable using passive components.

This filter may be better understood if it is seen as consisting of a capacitor in series with a simulated inductor, as in Fig. 7-5. For component values given, the simulated inductor represents an approximate 100H inductance with a Q of about 0.6. Needless to say, that kind of a coil isn't lying around in everybody's junkbox!

HIGH-PASS AMPLIFIER WITH HIGHER Q

The filter of Fig. 7-4B has the same form as the low-pass filter of Fig. 7-2B, but with resistors where there were

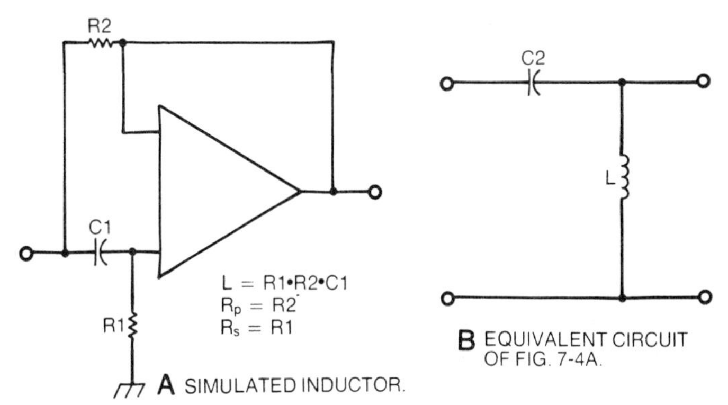

Fig. 7-5. An op-amp can be connected to simulate an inductor. R_p and R_s are the equivalent series and parallel resistances of the inductor.

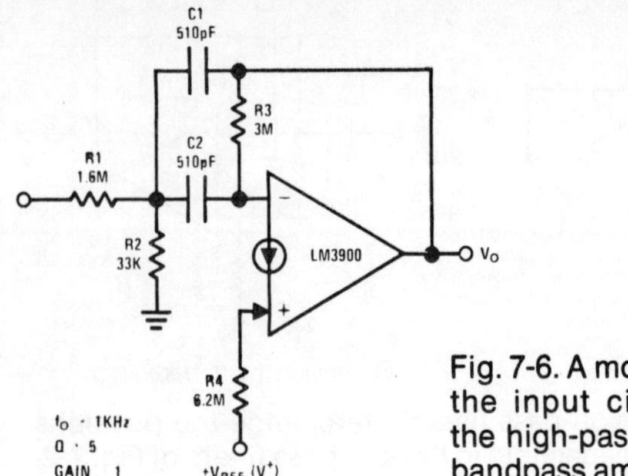

Fig. 7-6. A modest change in the input circuit changes the high-pass amplifier to a bandpass amplifier.

formerly capacitors, and vice versa. While the voltage-follower low- and high-pass filters of Figs. 7-2 and 7-4 use the same design equations, this is not possible in this design, since resistors outnumber capacitors in one case and capacitors outnumber resistors in the other.

In designing this high-pass filter, we start, as in the low-pass filter, with the requirements for gain, *A*, cutoff frequency, f_c, and Q that are imposed on us by our application. We next let C1 equal C3 and choose a starting capacitance for these components. Then,

$$R1 = \frac{1}{Q\ 2\pi f_c\ C1\ (2A + 1)}$$

$$R2 = \frac{Q}{2\pi f_c\ C1}\ (2A + 1)$$

$$C2 = C1/A$$

SINGLE-AMPLIFIER BANDPASS FILTER

A modest bandpass filter with relatively low gain, low passband frequency, and low Q (less than 10) can be realized using a single amplifier, as in Fig. 7-6. As in previous examples, we begin our design by determining what values of passband gain, *A*, passband center frequency, f_o, and Q are required. If we then let C1 and C2 equal the same arbitrary value, we can use the following equations for the rest of the components in the filter.

$$R1 = \frac{Q}{A\ 2\pi f_o\ C1}$$

$$R2 = \frac{Q}{(2Q^2 - A)\, 2\pi f_o\, C1}$$

$$R3 = \frac{2Q}{2\pi f_o\, C1}$$

$$R4 = 2 \times R3$$

HIGH-Q BANDPASS AMPLIFIER

Two amplifiers permit Q up to 50 in a bandpass filter (Fig. 7-7). This circuit resembles the previous bandpass filter, except that it adds a gain stage with positive feedback to boost Q. The design procedure starts with selection of Q and f_o. Then, a factor, k, is arbitrarily chosen. Typically, k lies between 1 and 10. A larger k will mean more gain, but it will also increase the spread among component values. Once k is decided upon, the other formulas are

$$R1 = R4 = R6 = \frac{Q}{2\pi f_o\, C1}$$

$$R2 = R1 \left(\frac{kQ}{2Q - 1}\right)$$

$$R3 = \frac{R1}{Q^2 - 1 - 2/k + 1/kQ}$$

$$R5 = 2 \times R4$$

$$R7 = k \times R1$$

$$R8 = 2 \left(\frac{R6 \times R7}{R6 + R7}\right)$$

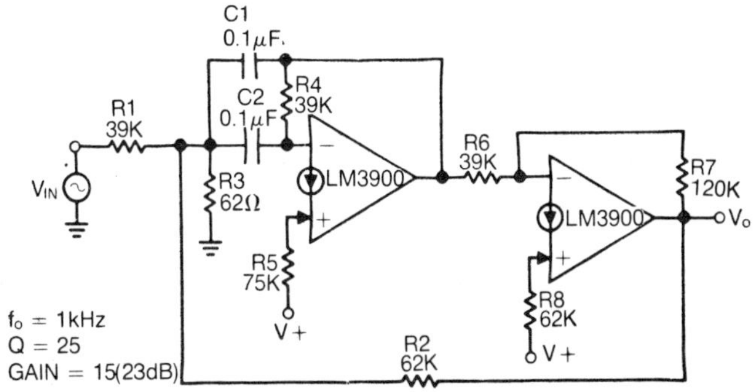

Fig. 7-7. Employing a separate amplifier for positive feedback boosts the Q of the preceding bandpass amplifier at the cost of three additional resistors.

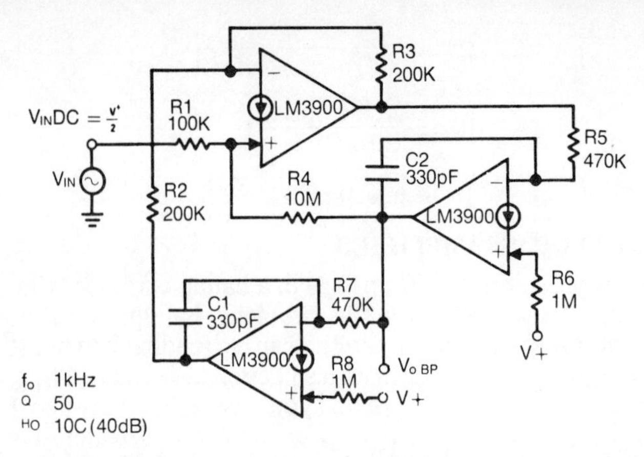

Fig. 7-8. Multiplying apparent capacitance by a process equivalent to Miller effect, two integrator-connected amplifiers help boost Q of this high-pass filter to more than 50.

BANDPASS AMPLIFIER WITH Q > 50

For a bandpass amplifier with a value of Q greater than 50, three amplifiers can be used, as in Fig. 7-8. This design has the additional advantage that its performance is less sensitive to element variation than was the performance of previous designs, and this simplifies the calculations. The additional amplifier represents an increase only in the number of passive components, since the LM3900 is, as you will remember, a group of four amplifiers in one IC. In designing such a filter, we once again start with the values of Q and f_o required by our application. Then, we let C1 equal C2, R2 and R3 equal 2R1, choosing arbitrary starting values for these components. The remaining resistance values are completed using

$$R4 = R1\,(2Q - 1)$$

$$R5 = R7 = \frac{1}{2\pi f_o\, C1}$$

$$R6 = R8 = 2\,R5$$

$$A = \frac{R4}{R1}$$

BANDPASS AND NOTCH IN A SINGLE FILTER

As Fig. 7-9 shows, the Motorola MC3401 is a quadraphonic amplifier array that is functionally and pin-for-pin

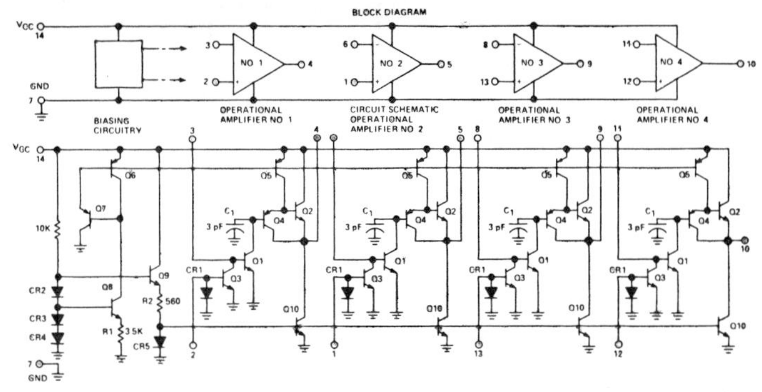

Fig. 7-9. The Motorola MC3401 is interchangeable with National Semiconductor's LM3900.

interchangeable with National Semiconductor's LM3900. Each amplifier uses a current input stage in which Q3 and CR1 form a current *mirror* such that, from the standpoint of the bias current through the base of Q1, a current *out of* the inverting input has the same effect as a current *into* the noninverting input, and vice versa. The output signal is a voltage derived from emitter follower Q2. Motorola suggests an interesting circuit (Fig. 7-10) that uses all four amplifiers of the MC3401 to produce a filter that can be used as either a bandpass or notch filter. The upper three amplifiers in the drawing form the bandpass filter. Amplifier 1 is a basic low-pass filter modified to a bandpass configuration by the capacitor in the input that effectively blocks DC and low frequencies. The high-frequency

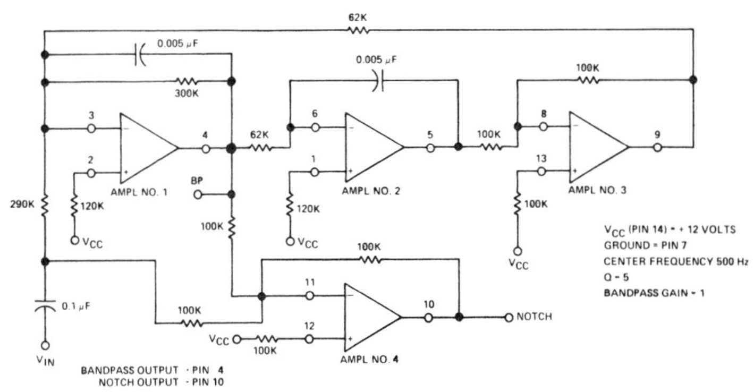

Fig. 7-10. All four amplifiers of an MC3401 or LM3900 can be used in this combination bandwidth/notch filter. The input signal is summed with the phase-inverted bandpass output at the input of amplifier 4 to produce the notch response.

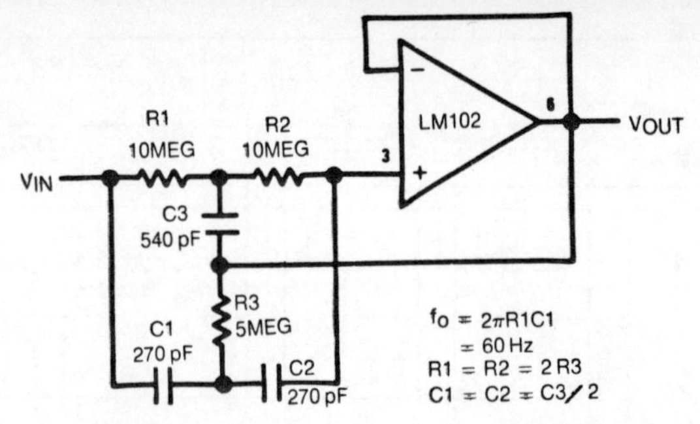

Fig. 7-11. Normally a low-Q resonant circuit, the twin-T can have its Q boosted to 50 by bootstrapping it to the output of a voltage follower.

rolloff of this filter is enhanced by the integrator (amplifier 2) and the inverter (amplifier 3) in the negative feedback loop. Since an integrator is essentially a low-pass filter with a cutoff frequency of zero, it adds to the rolloff of the first amplifier to produce a 40 dB/decade high-frequency rolloff. The inverter is needed to make sure the phase of the feedback signal is correct.

The most interesting feature of the filter is the way in which the bandpass response is made to produce a corresponding notch effect. The bandpass output, which is 180° out of phase with the input, is applied to a summing junction at amplifier 4, together with the input. Within the passband, the signals buck each other, and the output of the amplifier is attenuated. Outside the passband, the input signal has no competition and amplifier 4 operates with unity gain.

TWIN-T NOTCH FILTER WITH HIGH Q

Normally, the twin-T resonant circuit has a Q of about 0.3. By bootstrapping it to the output of a voltage follower, as in Fig. 7-11, Q can be increased to approximately 50, producing a very sharp notch filter. With the component values in the figure, notch frequency is 60 Hz. The high input impedance of the voltage follower makes it possible to use large values of resistance, with the advantage that capacitances can be small.

The high Q of the filter in Fig. 7-11 results in an extremely sharp notch. By adding another voltage follower as a variable gain stage, as in Fig. 7-12, we can vary Q from the 0.3 value of the basic twin-T to the 50 of the previous example. Changing Q does not affect either the center frequency or the depth of the notch, but it does vary the width of the notch, allowing more frequencies to fall into the filter's stop band.

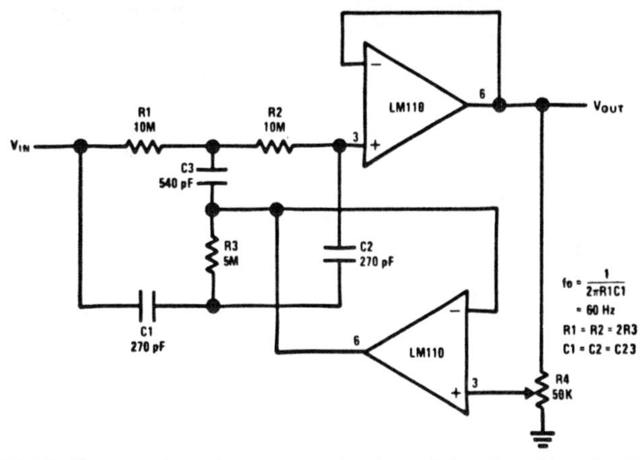

Fig. 7-12. By varying the amount of positive feedback, the Q of this twin-T notch filter and, consequently, its bandwidth can be varied widely.

HIGH FREQUENCY BANDPASS AMPLIFIER

The CA3015 in Fig. 7-13 is typical of RCA's basic op-amps, featuring a difference amplifier input stage with double-ended output driving an emitter follower through a difference

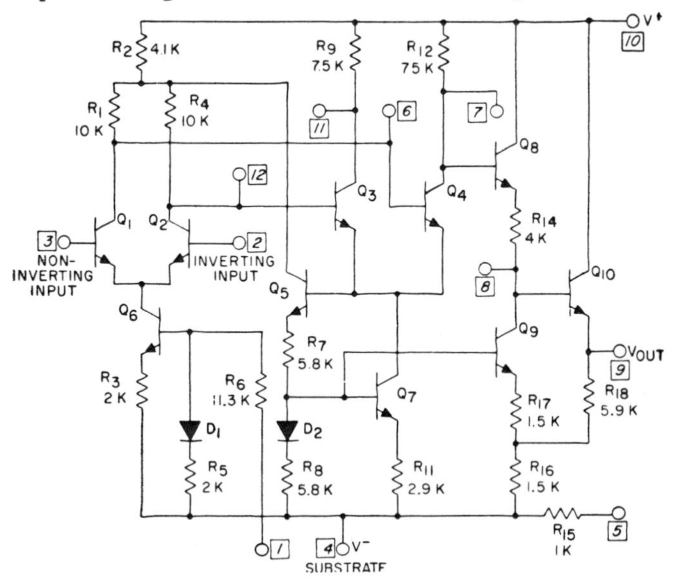

Fig. 7-13. The RCA CA3015 basic op-amp is a classic design. This figure identifies its pin assignments to help clarify the compensation applied in Fig. 7-14.

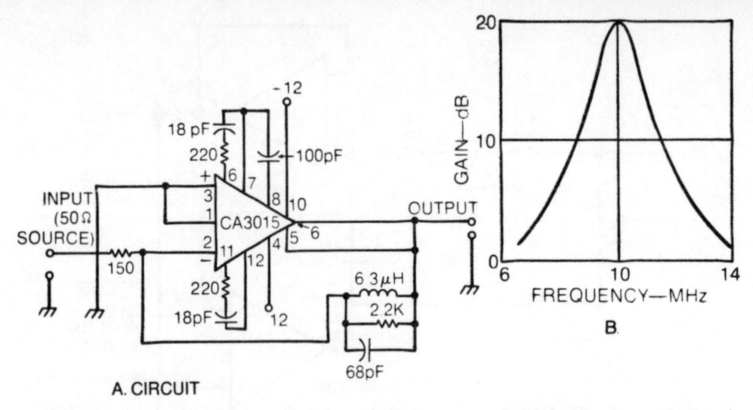

Fig. 7-14. With full compensation, the CA3015 operates at 10 MHz as a bandpass amplifier. In this circuit, Q is largely controlled by the resistor in the feedback loop.

amplifier buffer and level shifter. Normal supply voltage for the CA3015 is 12V. It is packaged in a 12-pin TO-5 can.

Figure 7-14 shows the CA3015, fully compensated for high-frequency operation, used as a 10 MHz bandpass amplifier. Its operation resembles the operation of the simple low- and high-pass amplifiers we considered at the start of the chapter. In this case, though, the feedback impedance is a parallel RLC network. A resistance is included in the network to lower the Q of the resonant circuit in order to increase the bandwidth. The component values in the figure are selected to provide a gain of 20 dB at midband and a Q of about 10. At 10 MHz, this gives a 3 dB (half-power) bandwidth of 1 MHz.

Q-MULTIPLIER

A circuit similar to the above is shown in Fig. 7-15. This uses a 741 op-amp with an LC network in the negative feedback loop, and positive feedback through R2 and R3. This is a Q multiplier intended for use in a shortwave receiver, so it is tuned to 455 kHz, corresponding to the receiver's IF. Depending on the coil used for L1, R1 will have to be selected by the cut-and-try method to make the filter gain 0 dB with the wiper of R3 at ground. Then, increasing positive feedback by sliding the wiper up R3 will increase Q, narrowing the IF selectivity and boosting gain. With a high-quality inductor for L1, augmented Q values of over 2000 can be obtained.

TUNABLE BANDPASS FILTER

A bandpass filter such as we encountered earlier (Fig. 7-6) can be tuned by varying the resistor designated R2 in that

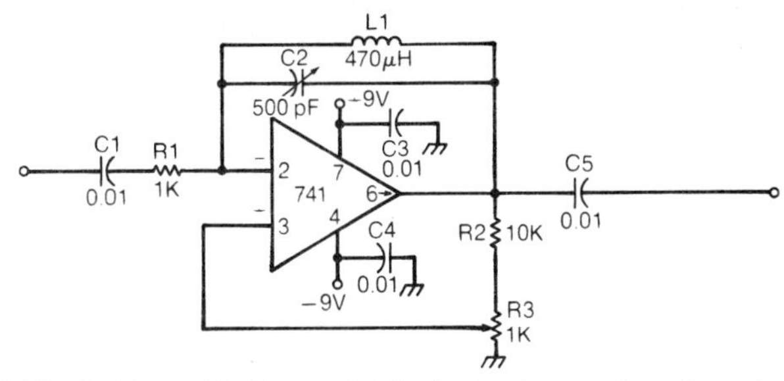

Fig. 7-15. A Q-multiplier, which is just another bandpass amplifier, albeit one with a much narrower passband than the previous example. Here, Q is controlled by varying the level of positive feedback. Capacitor C2 trims the Q multiplier to exactly 455 kHz.

diagram. An electrically tuned filter of this type is shown in Fig. 7-16. The right-hand op-amp is the filter itself, its R2 resistor replaced by an FET (Q1) used as a voltage-variable resistor.

A unique feature of this voltage-tuned filter is the control section, using the op-amp on the left of the diagram. This section linearizes the tuning of the filter. This is desirable,

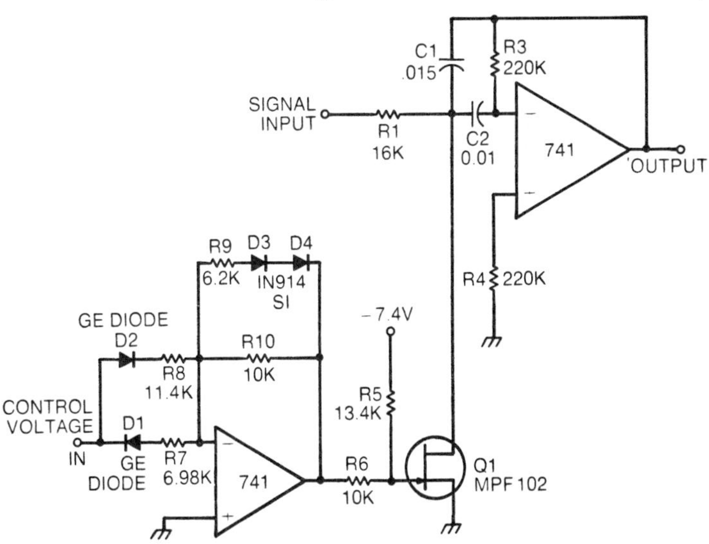

Fig. 7-16. Q1 is used as a voltage-variable resistor tuning the resonant frequency of A2 in this tunable bandpass filter. The diodes associated with A1 allow it to operate as a diode function generator, linearizing the response of Q1 to control voltage inputs.

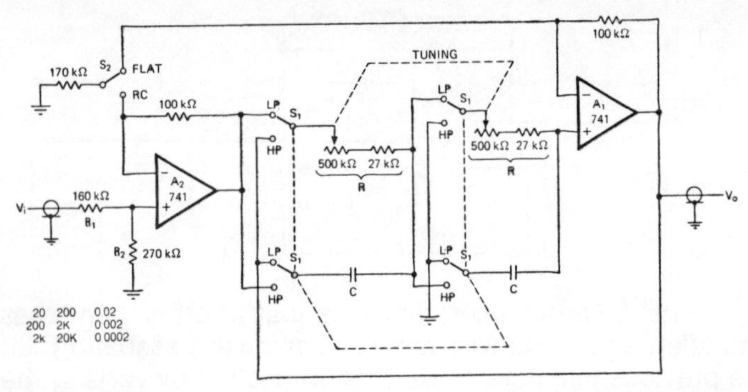

Fig. 7-17. A versatile filter with tunable cutoff frequency, switchable high- or low-pass response, and switchable RC or Butterworth characteristic.

since ordinarily the channel resistance of an FET does not vary linearly with gate voltage over a significant range. The op-amp circuit operates as a function generator that programs the FET to respond linearly to changes in input voltage. At zero control voltage, R5 and R6 bias the FET gate at −3V, setting the filter frequency, f_o, at 1460 Hz. When the control voltage is negative, D1 is the only diode to conduct; that point is determined by changing the FET (Q1) bias voltage through adjustment of the values of R10 and R7. If the control voltage is made positive, but less than 1.4V D2 alone conducts, and gain is set by R10 and R8. Above a positive control voltage of 1.4V (twice the 700 mV forward breakdown voltage of the 1N914s), D3 and D4 also conduct, reducing gain and maintaining a linear relationship between control voltage and Q1 channel resistance. This arrangement permits control voltages from +4 to −3V to tune the filter from 2640 to 570 Hz with excellent linearity. Diodes D1 and D2 are gold-doped germanium diodes for low forward breakdown voltage.

VERSATILE FILTER WITH SWITCHABLE RESPONSE

The relatively simple appearing filter in Fig. 7-17 has a number of interesting features. It is tunable over a frequency decade, it can be switched for either high or low pass, and it can be further switched to select either Bessel (RC) or Butterworth (maximally flat) response. Since the switches make the drawing somewhat complicated, the high- and low-pass versions of the filter are redrawn in Fig. 7-18. This makes it clear that these are basically two-section RC filters with one section bootstrapped to the output of their second op-amps.

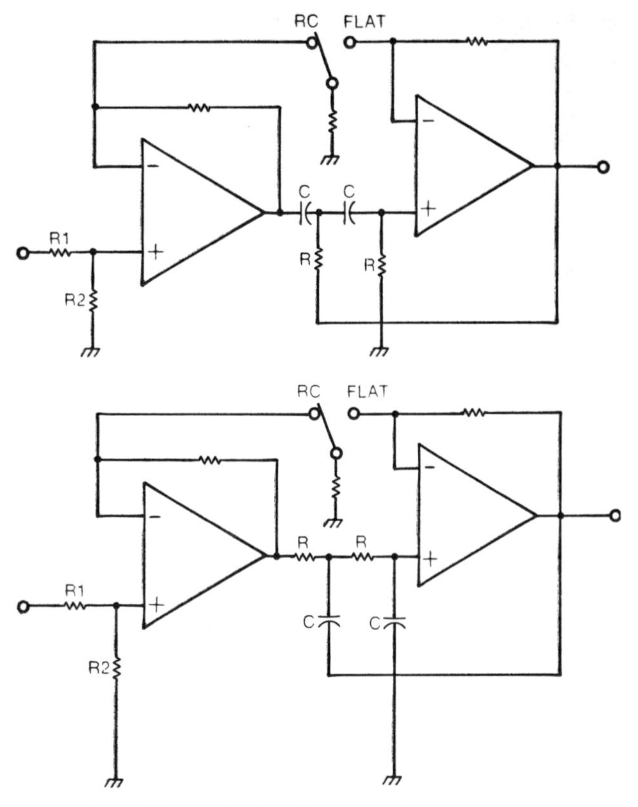

Fig. 7-18. The filter of Fig. 7-17, redrawn in its high- and low-pass modes to make its operation more clear.

Special Purpose ICs and Applications

The voltage regulators, op-amps, and comparators we have seen in the earlier chapters of this book are "building-block" linear ICs. They can be interconnected in a variety of ways to perform a variety of functions. There is a class of linear ICs that includes those devices designed to perform unique functions. We have already seen a few of the simpler ones: National Semiconductor's LM170 IF amplifier with AGC, and RCA's CA3123E AM receiver subsystem. In this chapter, we encounter some more complicated functional block ICs to see how they are implemented from simpler building blocks. In addition, this chapter includes some applications of building-block ICs that do not fall readily under any other chapter heading.

DIGITAL/ANALOG CONVERSION

Modern data processing methods create a dichotomy in our lives. Most of the data that we deal with is analog in the sense that it is continuously variable. Consider some examples of analog data that we might wish to subject to computer processing: the body temperature of an astronaut, the static air pressure around an aircraft in flight, the head and feed velocities of an automated milling machine in a factory, to name but a few. If we wish to use our most sophisticated processing methods on this data, we must interpret and code it in the binary format used by digital computers. This process of interpreting and encoding is called analog-to-digital conversion, and the devices that do the job are called A-to-D converters, or ADCs.

Once the data has been processed by a digital computer, we may or may not be satisfied to leave it in digital form. In terms of human perception, digital displays are good when the data presented is unchanging. It is much easier to read a digital voltmeter to five-figure precision that it is to read even a finely graduated, mirror-backed d'Arsonval voltmeter to the same precision. But is the reading changing? Is it going up or down? *Human factors* engineers know that a meter-type display is much easier for people to interpret in terms of rate of change than is a digital display. In terms of control systems like the one directing the work of that factory milling machine, it is necessary to have an analog signal if the computer is to change the head and feed speeds. So we see that after data has been digitally processed, it is often necessary to convert it back to an analog form. This is digital-to-analog conversion, and the devices that do this are D-to-A converters, or DACs.

DACs and ADCs require a combination of digital and linear integrated circuitry. Since the ADC is the more complicated, and since it requires a DAC as part of itself, we will look at the DAC first.

A Simple 4-Bit DAC

The DAC in Fig. 8-1 is designed to handle 4-bit binary words. Its output voltage varies between approximately 0.4

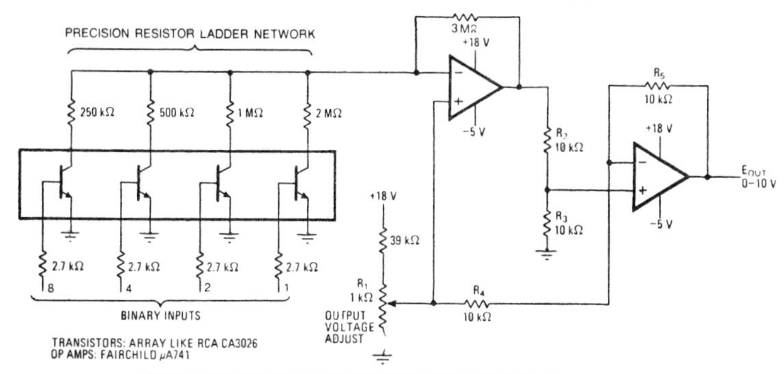

Input Word Decimal	Binary	Ladder Resistance	Gain	Output (V_{ref} = 0.425V)
0	0000	0	1	0.425
1	0001	2M	2.5	1.06
2	0010	1M	4.0	1.70
3	0011	667K	5.5	2.34
4	0100	500K	7.0	2.97
5	0101	400K	8.5	3.61
6	0110	333K	10.0	4.25
7	0111	286K	11.5	4.80
8	1000	250K	13.0	5.53
9	1001	222K	14.5	6.16
10	1010	200K	16.0	6.80
11	1011	182K	17.5	7.44
12	1100	167K	19.0	8.08
13	1101	154K	20.5	8.72
14	1110	143K	22.0	9.35
15	1111	133K	23.5	9.99

Fig. 8-1. In this digital-to-analog converter, the 4-bit parallel binary word is programed via the transistor switches and the analog output appears as a voltage. The table explains operation.

and 10V, depending on the word programed into it. This is controlled by the switches and the resistive-ladder network on the left side of the drawing. A *high* logic signal fed into the base of any transistor turns that transistor on and connects the selected ladder resistor from the inverting input of the op-amp to ground. With four transistors, there are sixteen possible resistance values that may appear at the amplifier inverting input. The ladder resistances are selected so that the gain of the op-amp increases in uniform steps for each unit increase in the binary input. Then, the reference voltage applied to the noninverting input of the first op-amp is selected by adjusting R1 so that the DAC output (10V) corresponds to an input of 15 (binary 1111). The second op-amp in the figure is used to minimize offset error.

This simple 4-bit DAC can be expanded to a 12-bit capacity by increasing the number of transistor switches and ladder resistors and changing the value of the feedback resistor in the first op-amp.

ADC Operation

The basic design for an ADC is shown in Fig. 8-2. The output of the up/down counter represents a certain quantity, expressed in binary form. This quantity is converted to its analog equivalent by the DAC at the bottom of the drawing. This analog representation of the counter output is compared to the analog input signal by comparator A1. If the two signals correspond, nothing further happens. If there is a discrepancy, the comparator switches to either its *high* or its *low* state,

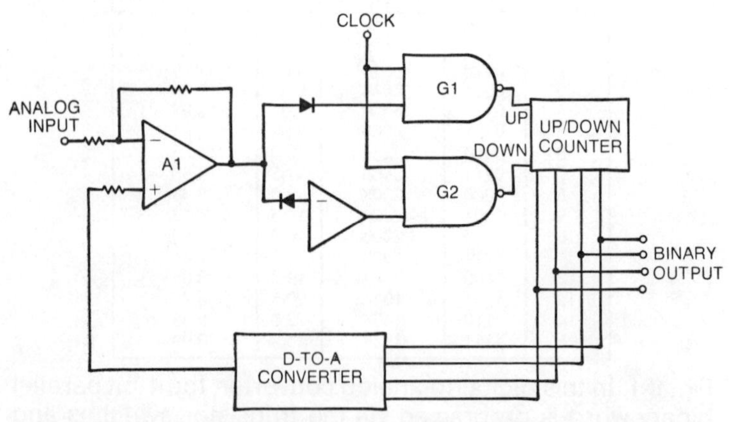

Fig. 8-2. This generalized analog-to-digital converter diagram gives an indication of the complexity of practical ADCs.

depending on the sign of the error. The comparator output applies a signal to either the up- or down-counting NOR gate, depending on whether it is high or low. The selected NOR gate then applies the clock signal to the counter, causing it to count up or down until the DAC output matches the analog input voltage.

A Monolithic IC DAC

Motorola's MC1508L-8 (Fig. 8-3) is a complete monolithic DAC on a 16-pin DIP. It differs from the DAC in Fig. 8-1 in that it has current inputs and output. In addition to its greater capacity, the MC1508L-8 is also a faster device, due to the sophistication of its current switches and its lack of an op-amp output.

Digital Input Controls Meter Reading

One simple application of the MC1508L-8 is the digital-input, meter-reading-output circuit in Fig. 8-4. The current from the DAC depends on V_{ref} and R14, so full scale on the meter should correspond to $V_{ref}/R14$. These two variables should be chosen so that the maximum current is less than 2 mA, the maximum allowable output current from the DAC. Resistor R15 should equal R14. A satisfactory value for compensation capacitor C is on the order of 50 pF.

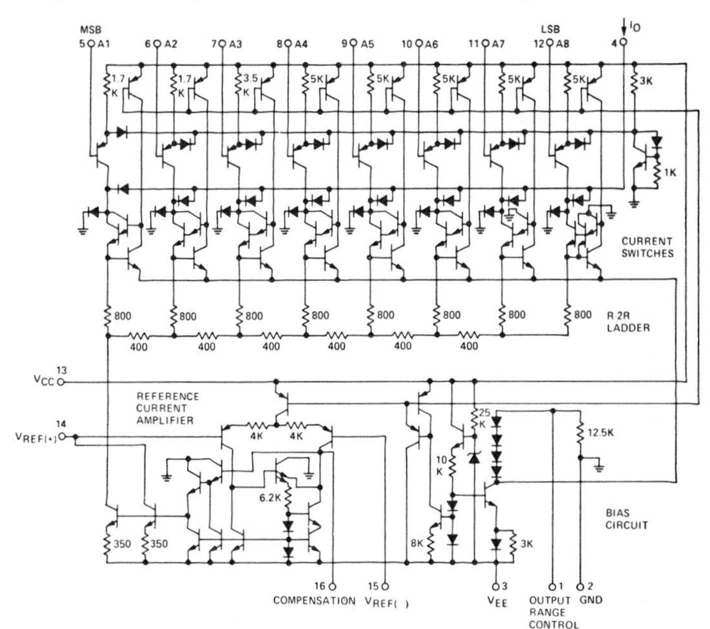

Fig. 8-3. Motorola's MC1508L-8 monolithic IC 8-bit DAC.

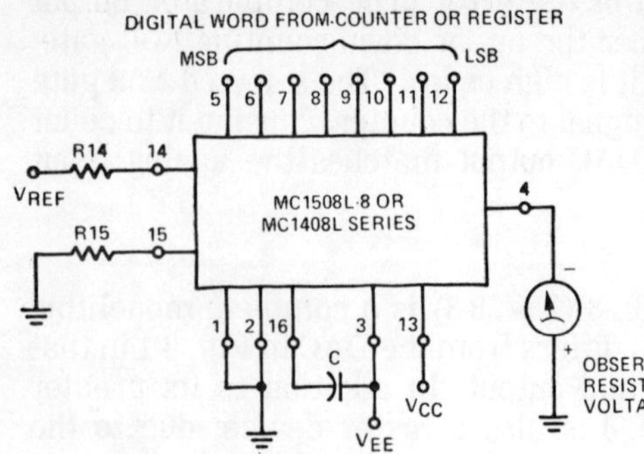

Fig. 8-4. In a simple application, the MC1508L-8 converts an 8-bit binary word to a meter reading.

A Linear IC ADC System

Motorola also makes an ADC control circuit, the MC1507, which is intended to work with the MC1508L-8 and Motorola MC74193 up/down counters to provide a high-speed tracking ADC as shown in Fig. 8-5.

FOUR-CHANNEL STEREO

The most significant innovation to occur in the high fidelity sound industry in recent years has been four-channel stereo, a sound reproduction process in which the customary two stereo speakers in front of the listener are augmented by two additional stereo speakers behind him (Fig. 8-6). With each speaker driven by its own audio signal, two advantages are claimed. The first is that it is no longer necessary to sit on a line directly between two speakers to obtain a realistic stereo effect, and the second is that, by overcoming room echoes with the rear speakers, true concert hall sound can be achieved in any room.

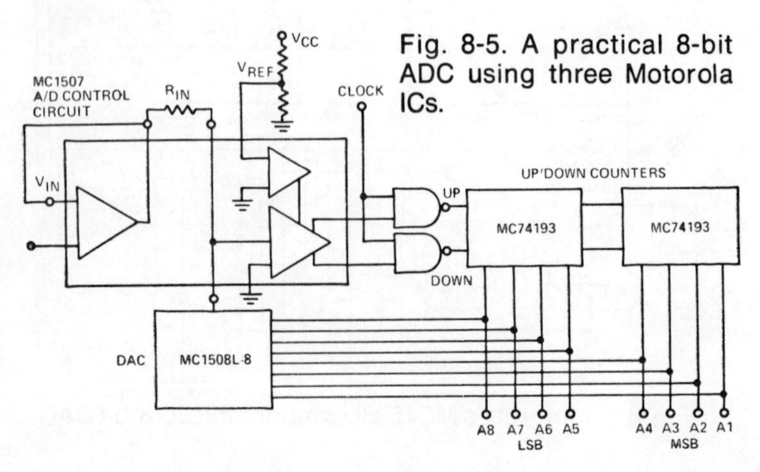

Fig. 8-5. A practical 8-bit ADC using three Motorola ICs.

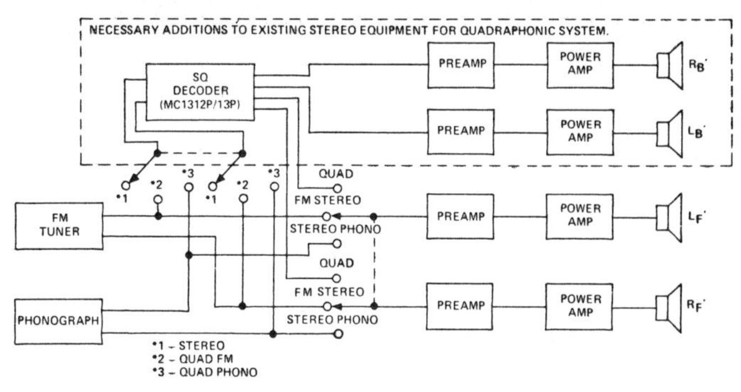

Fig. 8-6. Four-channel stereo adds extra dimension (and expense) with a decoder and two additional channels of audio.

There are several approaches to four-channel stereo, but presently, the trademarked CBS SQ system seems most likely to take over the home entertainment market. This is because the CBS SQ system is most compatible with conventional stereo equipment. SQ signals can be recorded on phonograph discs, two-channel tapes, and they can be broadcast by FM stations using conventional stereo multiplex equipment. Listeners with conventional stereo players and receivers will hear a full four-channel program.

SQ Encoding

All four-channel processes provide the normal right-front (RF) and left-front (LF) channels of conventional stereo. In addition, they provide a right-back (RB) and a left-back (LB) signal for the rear speakers. In the CBS SQ system, these four signals are encoded onto two channels, the right-total (RT) and left-total (LT). The two channels can be separated into four signals by using the phase relationships among the signals.

Stated mathematically, the left and right total signals are

$$LT = LF + 0.707\,RB - j0.707\,LB$$
$$RT = RF - 0.707\,LB + j0.707\,RB$$

The j in the expressions is a mathematical operator, but it can be understood as meaning that the phases of the signals operated on by j either lead or lag their actual phases by $\pm 90°$, depending on the sign of j.

An SQ Matrix Decoder

The LT and RT signals can be decoded by means of a matrix decoder like the one in Fig. 8-7. The decoder contains

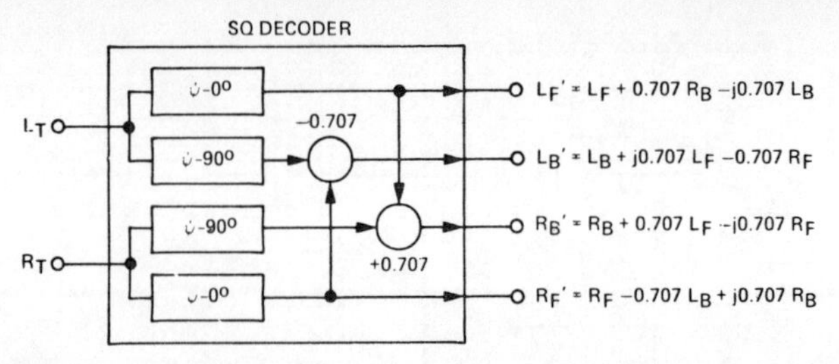

Fig. 8-7. As the figure shows, the four audio channels recovered by a matrix decoder in the CBS SQ system are not exactly the same as the original four channels on the master tape at the recording studio. The channels are marked with primes in this and in following figures to emphasize this.

broadband 90° phase shift networks and summing attenuators. As the figure indicates, the recovered signals are not exactly identical to the RF, LF, RB, and LB signals that the recording engineer started out with. However, for many low-budget applications, the signals that are obtained provide satisfactory four-channel sound. If it is desired to recreate more precisely the original four channels, additional electronic logic can be added.

Motorola's MC1312P and MC1313P are SQ decoders such as we have described. Both are sold in 14-pin DIPs, but the MC1312P is designed for 20V home entertainment power supplies, and the MC1313 is designed to run from an automobile's 12V system. Figure 8-8 shows how these SQ

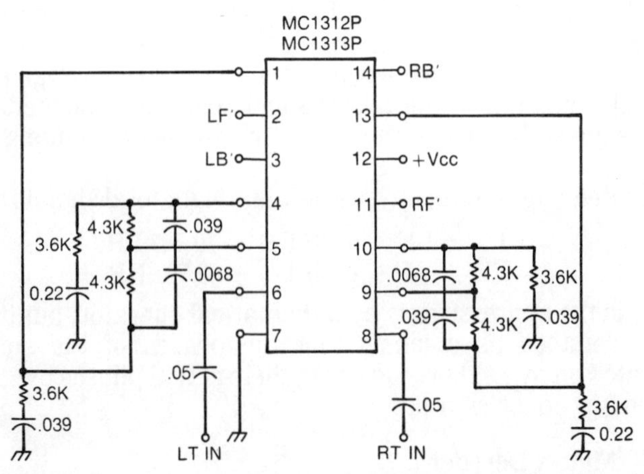

Fig. 8-8. A practical application of Motorola's MC1312/1313 SQ decoders.

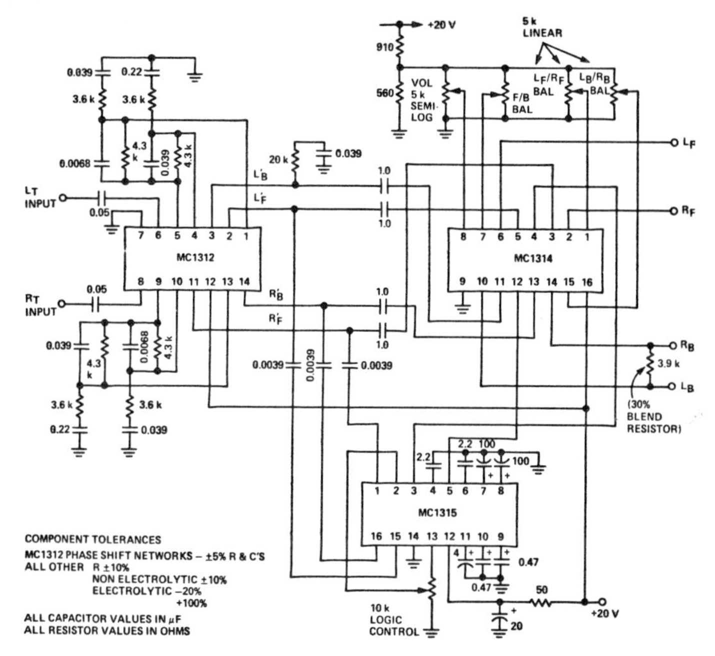

Fig. 8-9. This system uses additional IC logic to derive the original four channels from the output of the MC1313.

decoders can be used in a simple home entertainment or auto system. The more complex system of Fig. 8-9 uses additional Motorola logic devices to achieve a system that reconstitutes exactly the original four channels.

A LONG-PERIOD TIMER

Still one more application for RCA's versatile CA3094 programmable power switch (Fig. 2-23) is the timer in Fig. 8-10. Timing begins when the START switch is momentarily

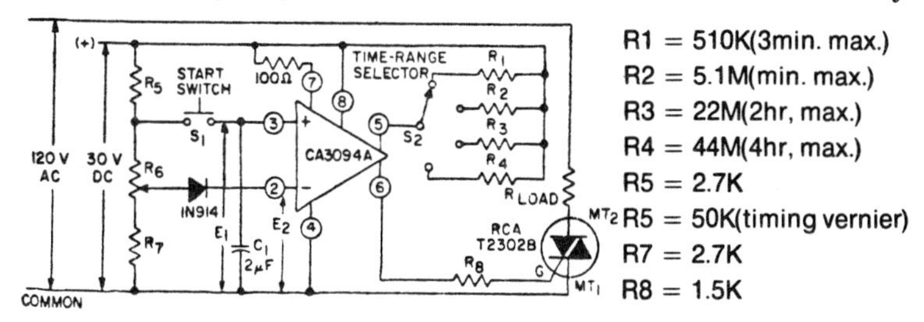

Fig. 8-10. A versatile timer. Timing range is set by switching in R1 through R4; duration within that time period is set by R6.

depressed, charging capacitor C1 through R5. Since the output of the CA3094 is now high, the triac turns on and power is applied to the load. When the start switch is released, the capacitor begins to discharge through the IC at a rate determined by the bias on pin 5, which is set by one of the time-range resistors. The capacitor discharges until the voltage on pin 3 is less than the voltage on pin 2, at which point the CA3094 output latches to its low state, removing the gate bias from the triac, which shuts off within a half-cycle of line voltage. By varying the voltage on pin 2 by means of R6, we can fine-tune the time period from zero up to the maximum time indicated in the figure. Resistor R7 insures that there is a minimum potential of 1V on pin 2, even with the wiper of R6 at the maximum time setting. The diode limits the maximum differential-input voltage to 5V.

A HIGH-PERFORMANCE AMPLITUDE MODULATOR

The RCA CA3080A, which, you will recall, resembles a CA3094 without the power output stage, can be used as an amplitude modulator, as shown in Fig. 8-11. The modulating signal could be applied directly to pin 5, but by applying the bias from a current source (the 2N4037 transistor), greater linearity is achieved. Performance is improved further by driving the 2N4037 with a constant-current source-regulated emitter follower, as shown. This helps to minimize temperature and supply-voltage effects.

ONE-WATT RF AMP IS GOOD TO 100 MHZ

Given a modulator, what more could we ask than a power amplifier with which to couple it? The National Semiconductor LF0063 buffer amplifier can be used as a power amplifier at frequencies up to 100 MHz. The circuit in Fig. 8-12 is straightforward, with a pi-network tank in the input and no

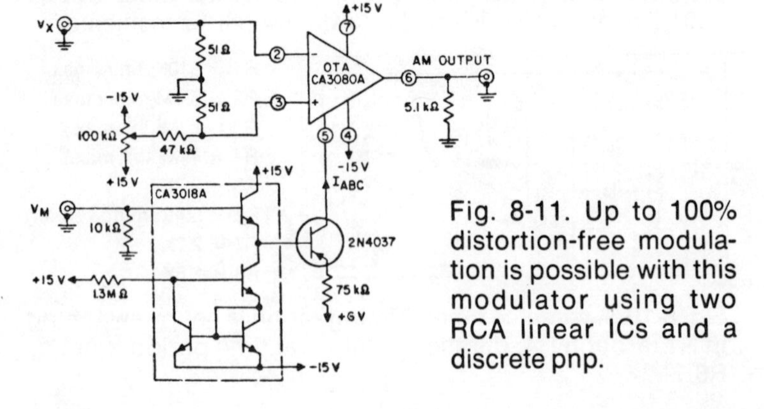

Fig. 8-11. Up to 100% distortion-free modulation is possible with this modulator using two RCA linear ICs and a discrete pnp.

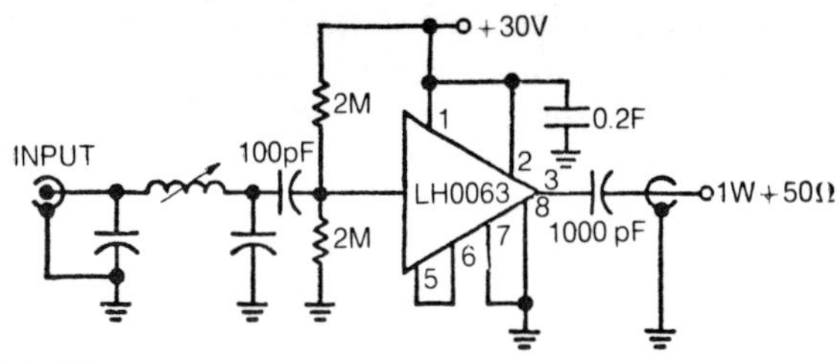

Fig. 8-12. The low-impedance ouput of the LH0063 can drive a 50Ω transmission line directly. Maximum CW power output is 1W with proper heatsinking of the LH0063's TO-3 package.

matching required in the output. The LH0063 comes in an 8-pin TO-3 package. The package is electrically isolated from the IC, so you can (and should) use it for shielding, especially at high frequencies. Pins *5* and *6* are for offset balancing, which is precalibrated for a 12 mV maximum with these two pins shorted together.

AUTOMOBILE TACHOMETER FOR CD IGNITIONS

Automobile tachometer circuits that are designed for standard Kettering ignitions often do not work well when used with capacitive-discharge (CD) ignition systems. The tachometer in Fig. 8-13 is intended to operate using the 14V pulses that are developed across the breaker points in a CD ignition. The circuit cannot be fooled by breaker point bounce.

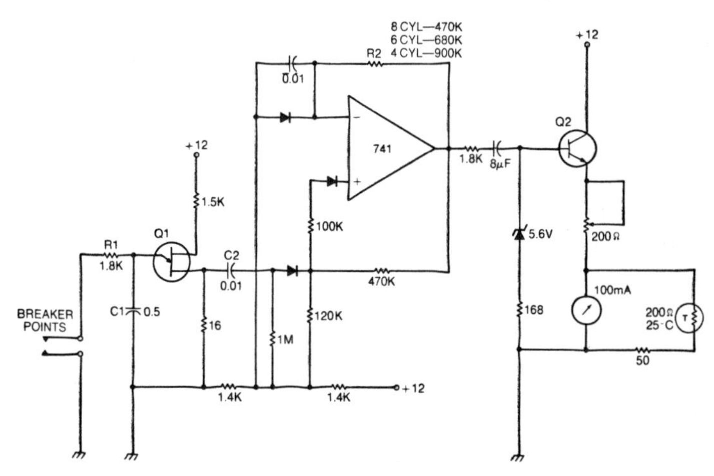

Fig. 8-13. You say your auto tachometer doesn't work since you put in a capacitive discharge ignition? Replace it with this tachometer and put the snap back in your driving.

The tachometer consists of three sections: a relaxation oscillator to suppress point bounce, a monostable multivibrator to generate a series of standard pulses in time with the distributor point openings, and a temperature- and voltage-compensated buffer for driving a meter.

Each time the distributor points open, capacitor C1 charges through R1 until unijunction Q1 fires. As long as the points are open, the current through R1 keeps Q1 on and prevents C1 from charging. The capacitor is large enough so that even if the points were to bounce on closing, they would not be open long enough to retrigger Q1.

Each pulse from Q1 triggers the op-amp in the monostable multivibrator to its saturated high state. It remains in this state until C2 charges to a high enough voltage to reset the op-amp comparator to its low state. Diode D2 clamps the voltage across C2 to about −0.7V, while D3 compensates for temperature changes that affect this clamp voltage. Since both diodes must be at the same temperature, they should be in physical contact with each other.

The series of uniform pulses from the monostable multivibrator triggers the base of Q2, varying the average current that flows through the meter. The physical inertia of the meter movement damps out the pulsations and presents an average reading. The output is regulated against supply-voltage fluctuations by the 5.6V zener and against temperature variations in Q2 by the thermistor.

ICE-WARNING INDICATOR

Three of the four amplifiers of an inexpensive LM3900, a thermistor, an LED, and a handful of resistors and capacitors can be used (Fig. 8-14) to make an ice-warning indicator that can advise drivers of possibly icy road conditions. The thermistor is mounted outside the car, in a location where it is out of the direct flow of air. When the outside temperature drops to 36°F, the LED begins flashing at a once per second rate. As the temperature drops below 32°, the LED turns on continuously. Amplifier A1 compares the current through the thermistor with the current through R1 and R2 and generates a voltage proportional to their difference. The constant of proportionality is controlled by R3. Amplifier A2 is a free-running multivibrator with a frequency of about 1 Hz. Its operating principle is the same as that of the simple square-wave oscillators of Chapter 3. Amplifier A3 compares the outputs of A1 and A2. Above 36°, the output of A1 is always less than the output of A2, and no current is supplied to the LED. In the temperature range from 36° to 32° the output of A1

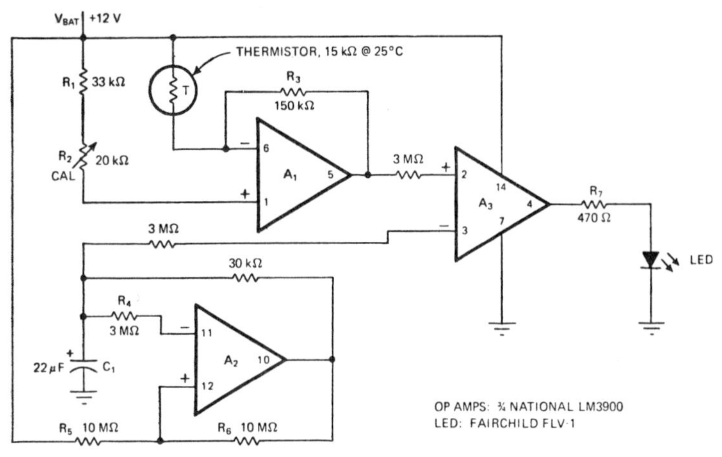

Fig. 8-14. Stealing an idea from the manufacturers of the Rover automobile, this project warns of impending icy driving conditions. Since it uses LM3900 "Norton" amplifiers to compare currents, it is insensitive to battery voltage changes.

is between 0.3 and 0.6 times the battery voltage, which is the range of the square-wave output of A2. This causes current to flow to the LED each time the multivibrator output goes low. Below 32°, the output of A1 is greater than 0.6 times the battery voltage, and the LED stays lighted continuously. Since the sensing circuit is sensitive to a current ratio and not to an absolute voltage, the calibration of the ice alarm is not affected by battery voltage variations.

To calibrate the circuit, immerse the thermistor in a mixture of melting ice and water and adjust resistor R2 just to the point where the LED stays on continuously.

THE CA3062 LIGHT-CONTROLLED POWER SWITCH

RCA's CA3062 is a unique IC that incorporates a dual photo-darlington pair and an op-amp power amplifier (Fig. 8-15) in a single TO-5 12-pin IC package with a transparent top. The photo-darlingtons turn on in the presence of light of intensity greater than approximately 50 lumens/ft^2 at the package window. This characteristic can be used to control the power amplifier so that it either turns on or off an external device. Transistor Q7 is the normally off switch, while transistor Q6 is normally on.

A Counting Control and an Alarm System

Two applications for the CA3062 are shown in Fig. 8-16. In the first, an output pulse is generated each time the CA3062

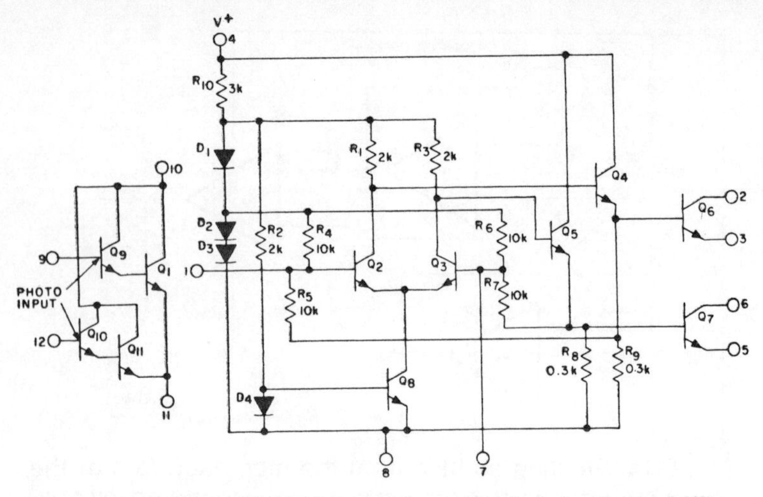

Fig. 8-15. RCA's CA3062 combines light-detecting ability and a power switch in a single TO-5 IC with a window in its top.

window is occluded by a passing object. The pulses can be used for counting. The second application uses both the normally on and the normally off outputs of the CA3062 in an alarm system. As long as nothing interrupts the light beam, power is supplied to both light and load, but when the light beam is interrupted, the triac on the left turns off, cutting power to light and load, and the triac on the right turns on, actuating the alarm.

ZERO-CROSSING DETECTORS

The normal technique for transforming a sine wave into a square wave is to amplify the sine wave and then clip the tops and bottoms from the amplified signal. For very slowly varying signals, it takes a lot of amplification before a square-sided output can be achieved this way. Since a fast rise and fall time is required for many counter-type circuits, a square-sided pulse is essential. A more efficient method for squaring a sine-wave signal than amplifying and clipping is shown in Fig. 8-17. It uses one comparator from a National Semiconductor LM139 quadrature comparator (Fig. 3-1) as a zero-crossing detector. The output of the detector changes state each time the input waveform passes through zero.

Two versions of the detector are shown in the figure. You can see that even though the LM139 is designed for single-supply operation, the detector circuit is simpler if a dual-voltage supply is used. In Fig. 8-17A, the detector, using

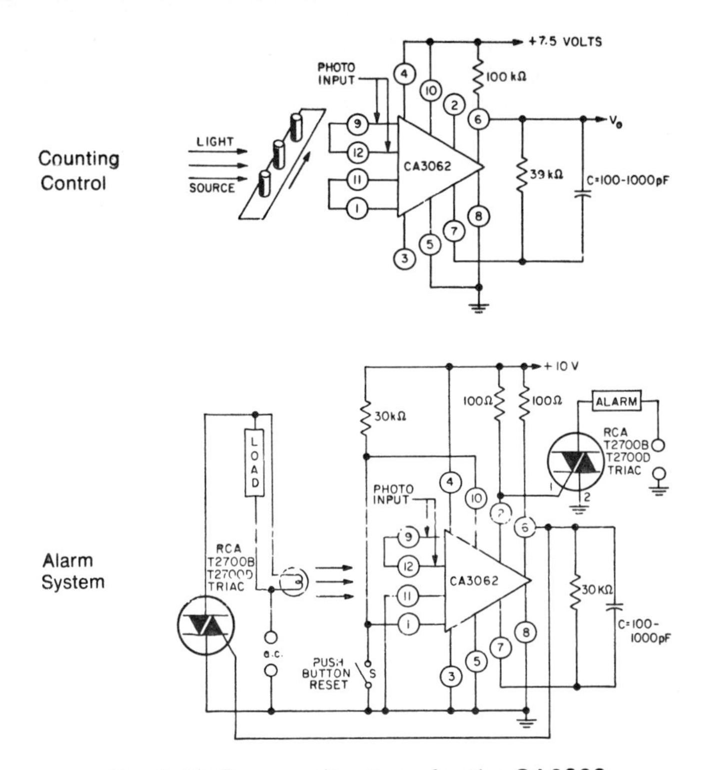

Fig. 8-16. Two applications for the CA3062.

positive feedback, compares the potential at its inverting input (V2) with the potential at its noninverting input (V1).

Assume that the input signal to the detector starts at zero volts and then rises. During the time that the signal is above zero in this first half-cycle, the comparator output remains at zero. As soon as the input voltage falls a little below zero on its second half-cycle, the comparator voltage immediately rises

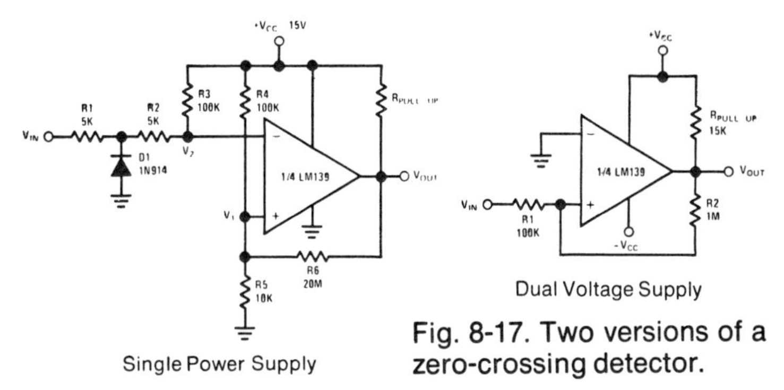

Fig. 8-17. Two versions of a zero-crossing detector.

to 15V and latches for the rest of the half-cycle. During this, and all other negative half-cycles, the voltage at the inverting input is held above zero by the clamping action of diode D1. When the signal again swings positive, the comparator output switches to zero as soon as the small hysteresis voltage is exceeded, and the output latches in that state until the signal again passes through zero.

The dual-supply zero-crossing detector (Fig. 8-17B) is simpler. Again, a small amount of hysteresis is used to assure latching, but the input signal is simply compared to the ground reference established at the inverting terminal. Each time the input signal exceeds the hysteresis voltage in the positive direction, the output latches in the high state. Each time it exceeds it in the negative direction, the comparator latches in the low state.

Zero-Crossing Switches and Controllers

The zero-crossing detector has more widespread application than merely squaring sine waves. RCA's CA3058 and CA3059 are ceramic and plastic 14-pin DIP versions of a zero-voltage switch. The function of such a switch is to delay

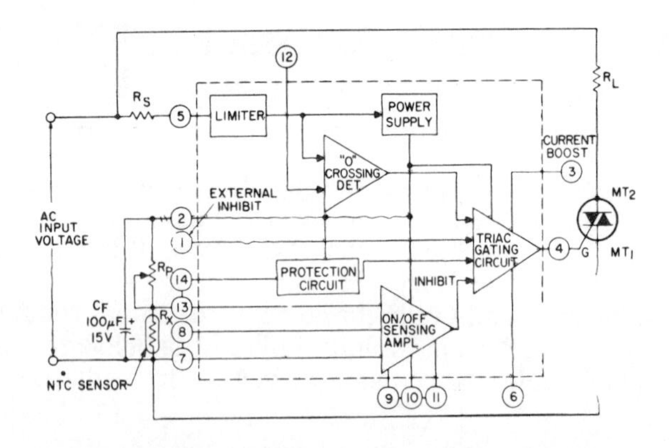

AC Input Voltage (50/60 or 400 Hz) V AC	Input Series Resistor (R_S) kΩ	Dissipation Rating for R_S W
24	2	0.5
120	10	2
208/230	20	4
277	25	5

Fig. 8-18. RCA's CA3058/3059 incorporates a zero-crossing detector to gate a triac only when the line voltage is passing through zero.

firing of a triac until the line voltage being controlled by the triac passes through zero during one of its AC cycles. This prevents back emf surges in inductive loads and triac-generated RF interference. The CA3058/3059 is shown in block diagram form in Fig. 8-18. It provides a number of functions: (1) The limiter/power supply permits the IC to be operated directly from the power-line voltage. (2) The on/off sensing amplifier is a difference amplifier with single-ended output that allows the IC to be controlled by external on/off signals. It can be used as a *comparator*, a *normally on* switch,

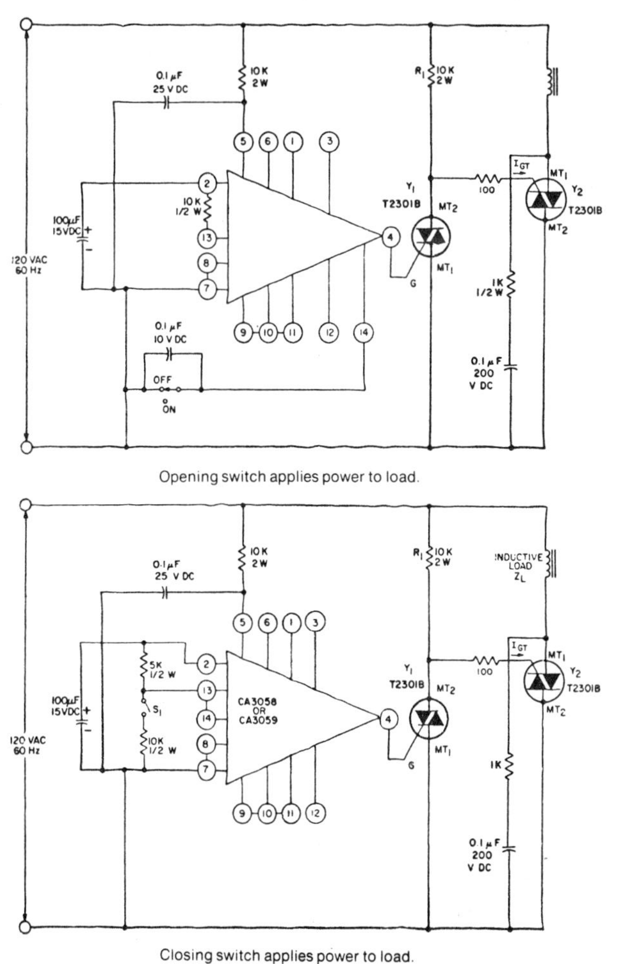

Fig. 8-19. Two ways of using a zero-voltage switch (ZVS). The choice of control method depends on the application and the kind of switch used.

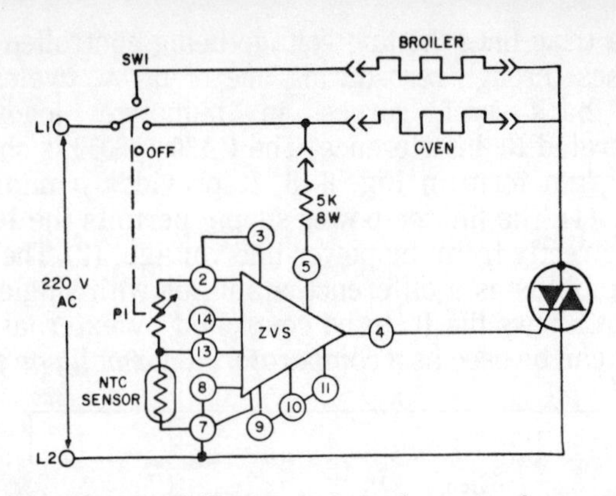

Fig. 8-20. Using a ZVS as an interference-free oven controller. Goodbye to those pops in the kitchen radio that used to accompany thermostat cycling! Fail-safe feature turns power off if thermostat breaks.

or a *normally off* switch. (3) The zero-crossing detector synchronizes the triac gating signal with the line voltage so that the triac fires when the line potential is zero. The external negative-temperature-coefficient sensor provision implements fail-safe operation by removing drive from the triac in the event the sensor opens or shorts.

Figure 8-19A and B shows two simple circuits using the CA3058/3059. In the first, power is applied to the load when the switch is opened. In the second, closing the switch applies power to the load. Both circuits use the external-sensor provision for on/off operation.

Figure 8-20 shows an oven-control circuit that uses the sensor provision to provide temperature control. The sensor compares the oven temperature with the required temperature set by potentiometer P1 and turns power to the heating elements off when the required temperature is reached. If the sensor fails, either by shorting or by opening, power to the heating elements is cut, assuring fail-safe operation.

Analog Computation

The operational amplifier, that ubiquitous linear building block, has quite an interesting history. Perhaps the first use of electronic amplifiers for a purpose other than communication or entertainment was in the M9 antiaircraft gun director developed during World War II by C.A. Lovell and D.B. Parkinson of Bell Laboratories. By selecting input and feedback impedances of their amplifiers, Lovell and Parkinson were able to create analog summers and integrators that could be interconnected to solve the differential equations representing the trajectories of aircraft and shells and make those trajectories intersect (sometimes!). After the war, J.R. Ragazzini, R.H. Randall, and F.A. Russel built a general-purpose electronic analog computer using op-amps. They described the amps in a paper in the May 1947 *Proceedings of the IRE*. This was the first *electronic* analog computer. (Vannevar Bush, at MIT, had built a mechanical differential analyzer before the war.)

These early operational amplifiers were large, complex, chopper-stabilized vacuum-tube DC amplifiers. They were bulky and hot and their failure rates were surpassed only by those of those early-day vacuum-tube *digital* computers. Nevertheless, in spite of their drawbacks, they opened up new territory in the study of differential equations and control systems. Commercial vacuum-tube analog computers were produced, and even as late as the mid-1960s, a few could be found in the engineering schools of most North American universities, still being called upon when the more powerful and reliable transistorized units were tied up.

The drive to produce an op-amp even more compact and reliable than the transistor units of the early 1960's grew out of engineers' experiences with earlier op-amps. They were finding more and more uses for them: in solving problems, in building control systems, and finally, even in communications and entertainment, when they began to see the potential for inductorless circuits. The Fairchild μA709 was probably the device that first made linear IC op-amps readily and

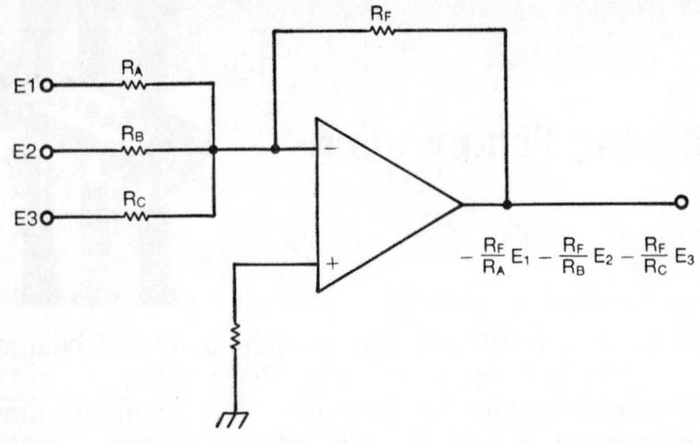

Fig. 9-1. A summer is basically an op-amp with negative feedback and several inputs. Its output is the inverted sum of the invidual inputs multiplied by their feedback to input resistance ratios.

inexpensively available to design engineers. The 709's popularity led to internally compensated op-amps like the 741, and eventually to the whole range of linear ICs that fills the earlier chapters of this book.

Since the analog computer op-amp is largely responsible for the growth and development of linear integrated circuit technology, it is fitting that in this final chapter we take a closer look at linear ICs as analog-computer elements.

The primary function of an analog computer is the solution of differential equations. The analog computer operator solves his equations by programing the computer so that the voltages at various points represent the variables being operated on by the differential equations. The computer becomes a time-scaled analog of a chemical process, a mechanical system, or even a living population.

The subject of differential equations requires a background in differential and integral calculus that not all readers of this book will have. Accordingly, we cannot go very deeply into the primary function of analog computers. There is, however, an area with which most readers are familiar, and in which certain analog-computer elements function in a way that can be readily understood. This area is that covered by ordinary algebraic equations.

SUMMERS

The basic arithmetic unit of the analog computer is the inverting summer (Fig. 9-1). This is just an op-amp with resistive feedback and several parallel inputs. If R_A, R_B, and

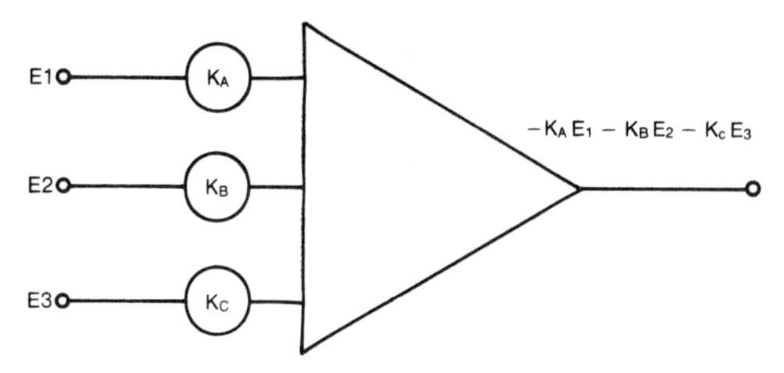

Fig. 9-2. The analog computer symbol for a summer with three inputs. Sign inversion is implicit in this symbol.

R_C are each equal to R_{fb}, the output voltage of the summer will be equal to the sum of the input voltages multiplied by −1. Practical analog-computer summers offer input potentiometers and selectable feedback resistors so that the summer output is equal to the product of some constant and the input voltage. Figure 9-2 shows the analog-computer symbol for a summer with constant-multiplying capability. The fact that it is an inverter is understood. The circles represent the pots, and the constants (K_A, K_B, etc.) represent the quantities by which the input voltages are multiplied.

Let's see how summers can be used to solve simultaneous linear algebraic equations. Take two simple linear equations:

$$y = 10x + 4 \qquad (1)$$

$$y = -6x + 3 \qquad (2)$$

Figure 9-3A shows how equation (1) can be represented using summers and a fixed −1V reference source. Within the limits of the summer supply voltages, any voltage applied to the x terminal will cause the voltage at the y terminal to represent the solution to equation (1) for that value of x. For example, if x is plus 0.5V, the potential at y will be plus 9V. Likewise, if x is −1V, the potential at y will be −6V.

Suppose we now rearrange equation (2) so that it is an expression for x in terms of y.

$$x = \frac{y}{6} + \frac{1}{2} \qquad (3)$$

Note that equations (2) and (3) are identical; the only change has been to interchange the dependent and the independent variables. A summer representation of (3) is shown in Fig. 9-3B. As in the previous example, any voltage input at y will cause a voltage output at x that corresponds to the value of x

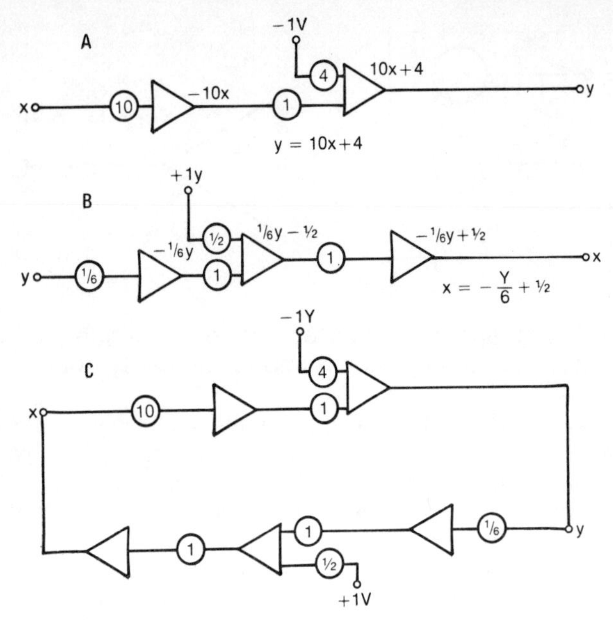

Fig. 9-3. The implementation of two linear algebraic equations and their simultaneous solution using analog computer summers.

for that value of y. If y is −6V, x will be + 1.5V. If y is + 18V, x will be −2.5V.

If we interconnect the two circuits, as in Fig. 9-3C, we force the circuit to represent the point of intersection—the simultaneous solution of the two equations. If we measure the potential at point y, it is 3.38V; at x, it is −0.062V.

Two practical summer-type circuits are shown in Fig. 9-4. The first is an inverting summer, such as we have been using. The second is a difference amplifier in which the sign of only one of the input signals is changed (V2). Note that with this circuit, the coefficient of the noninverted term is represented by a more complicated function of resistance than is the coefficient of the inverted term.

MULTIPLIERS

Multiplication of two or more variables is a little more complicated than is simple addition. The earliest analog computers used servo resolvers in which servomechanisms drove potentiometers in order to multiply two quantities. Modern multipliers are all electronic and a good deal more sophisticated. We encountered two functional-block

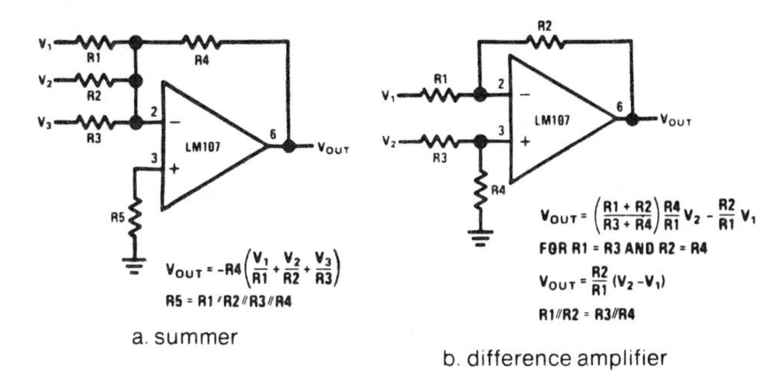

a. summer

b. difference amplifier

Fig. 9-4. Two practical linear IC summer circuits.

multipliers in Chapter 6. Let's see how multipliers can be used with basic building-block linear ICs.

RCA's CA3080 *operational transconductance amplifier* (OTA), with its ability to control transconductance by means of an external current signal, makes an excellent four-quadrant multiplier. There are several steps that must be taken to adjust the OTA multiplier of Fig. 9-5 for zero offset. These are listed below.

1. Set the 1M potentiometer on the output to the approximate center of its range.
2. Ground the x and y inputs and adjust the 100K potentiometer on the noninverting input until the output voltage is zero.
3. Ground the y input and apply an audio-frequency signal from a low-impedance generator to the x input. Adjust the 20K potentiometer in the transistor emitter lead to return the multiplier output voltage to zero.
4. Ground the x input and apply the test signal to the y input. Adjust the 1M potentiometer at the output for zero output voltage.

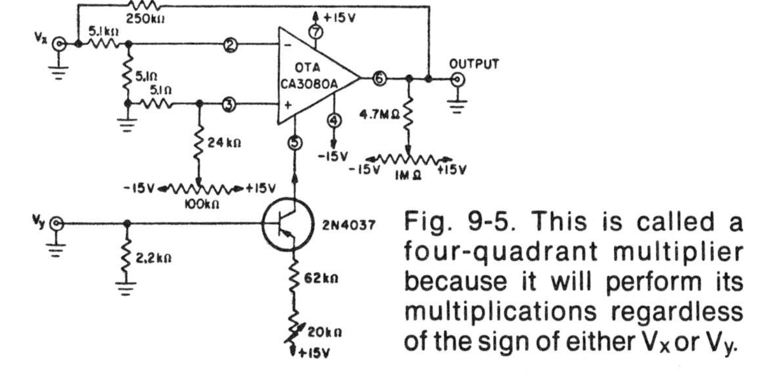

Fig. 9-5. This is called a four-quadrant multiplier because it will perform its multiplications regardless of the sign of either V_x or V_y.

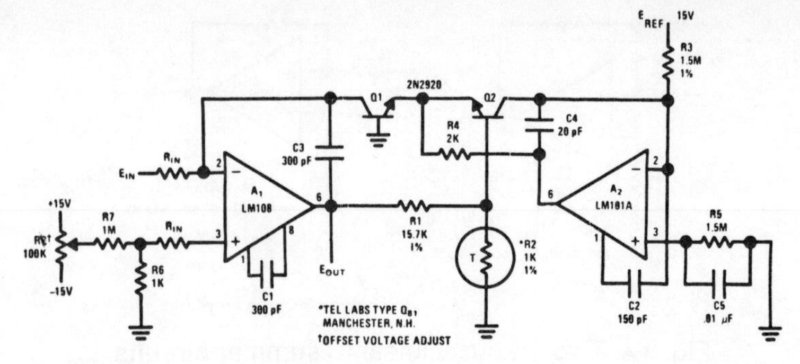

Fig. 9-6. A log generator with 100 dB dynamic range but somewhat narrow bandwidth.

Direct multiplication of two signals isn't the only way to implement a multiplier. We can also generate voltages that represent logarithms of our input signals, add these, and then generate a signal that is the antilog of the summation. If that sounds cumbersome for multiplication, it is. But logarithms open up a whole world of powers and roots to us. Remember that $x^{3.2}$ is just the antilog of 3.2 $(log\ x)$. The square root of x is just the antilog of ½ $(log\ x)$. Logarithms also give us the key to division, the quotient of a pair of numbers being simply the antilog of the difference of their logs. Where the OTA multiplier above cannot divide, a logarithmic system can.

But, you ask, aren't logs hard to generate? Not at all, thanks to the exactly logarithmic variation between collector current and emitter–base voltage in an ordinary bipolar transistor. In Fig. 9-6, Q1 is the feedback device that controls the log generator output response. The output of the circuit follows the relationship

$$E_{out} = -\left(\log_{10} \frac{E_{in}}{R_{in}} + 5\right) \quad (4)$$

That's not a mistake. The output *voltage* is equal to the log of the input *current* plus a constant. The log output is accurate to within 1% for any current between 40 and 400 mA. Outside that range, its accuracy is ±3% down to 10 nA and up to 1 mA.

Even with the frequency compensation used, the circuit in Fig. 9-6 is slow, requiring up to 5 msec for its output to settle to within 1% of its final value. A faster log generator is shown in Fig. 9-7. This is the same circuit redesigned to give greater speed but less range. It adds an LM102 voltage follower in the input to improve performance at low input-current levels.

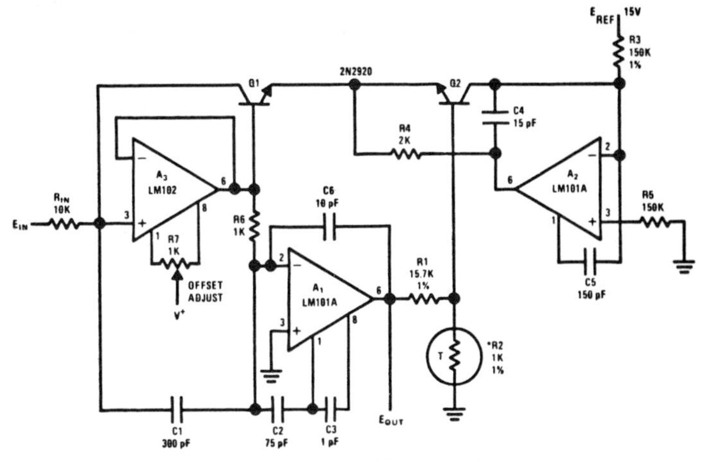

Fig. 9-7. This log generator is 100 times faster than the previous log generator, but its range is only 80 dB.

The antilog generator of Fig. 9-8 constitutes just a rearrangement of the circuit in Fig. 9-6. The combination of amplifier A1 and Q1 drives the emitter of Q2 in proportion to its input voltage. The collector current of Q2 varies exponentially with base–emitter voltage. This current is converted to a voltage by amplifier A2. The output voltage is

$$E_{out} = 10^{E_{in}} \tag{5}$$

Connecting the log generator of Fig. 9-6 with the antilog generator of Fig. 9-8 produces the one-quadrant multiplier of

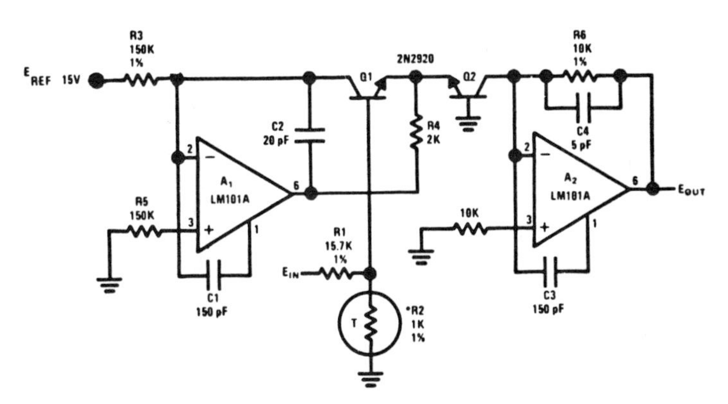

Fig. 9-8. A minor rearrangement of parts, and the log generator becomes an antilog generator (exponentiator).

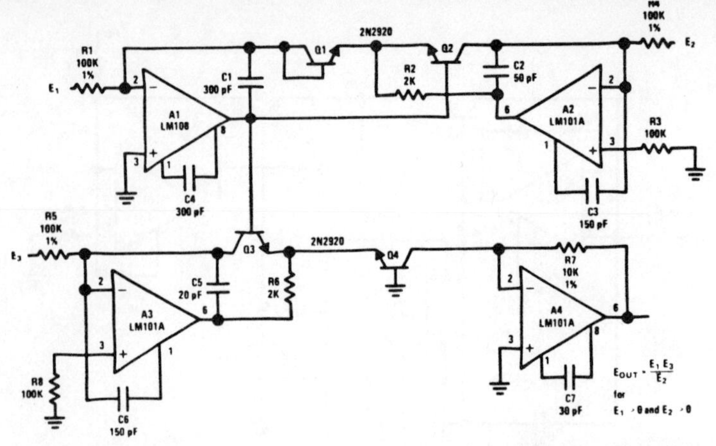

Fig. 9-9. Interconnecting the log and antilog generators produces a multiplier.

Fig. 9-9. The log generator output from A1 drives the base of Q3 with a voltage proportional to the log of E1/E2. To this is added a voltage proportional to the log of E3. and this total drives antilog transistor Q4. Amplifier A4 converts the collector current of Q4 to a voltage.

The cube generator in Fig. 9-10 is a variation on the multiplier in which two of the inputs are fixed. This circuit can

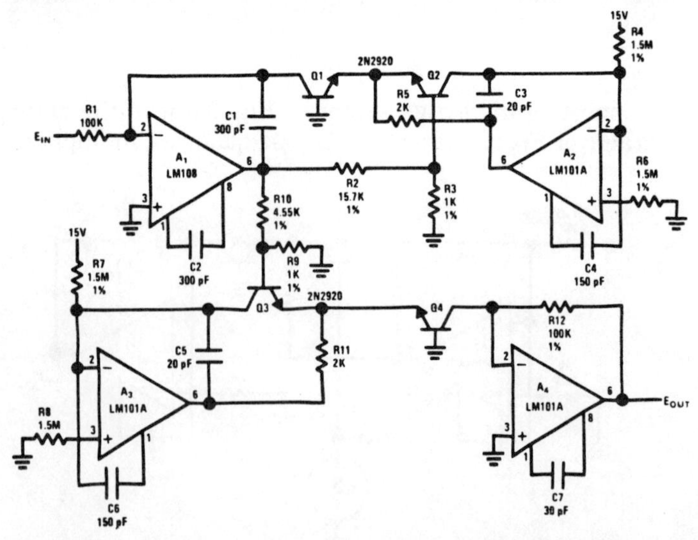

Fig. 9-10. When two inputs to the multiplier are held constant, it becomes capable of generating an infinity of powers of the input voltage by proper selection of R9 and R10.

generate a variety of powers, depending on the ratio of R9 and R10. The exact expression relating output to input voltage is

$$E_{out} = E_{in} \frac{16.7\,R9}{R9 + R10} \qquad (6)$$

The root extractor in Fig. 9-11 is based on similar principles. The log signal is produced by A1 and Q1, and this is level-shifted by Q2 to produce a zero signal at the voltage divider comprising R4 and R5 when the input signal is 1V. Adjusting R4 and R5 selects the root to be taken. When these resistors are equal in value as shown, the log signal is divided by two and the circuit produces square roots. If R4 were increased to 40K, division would be by three and a cube root output would result.

ANALOG COMPUTERS AND DIFFERENTIAL EQUATIONS

But what of the analog computer's primary application, the solution of differential equations? Approaching a subject like differential equations in a book like this presents some problems. This is not a book for engineers, but for ordinary hobbyists whose mathematic backgrounds may differ considerably over a wide range.

The material on the following pages constitutes a quick "cram" course in higher mathematics, followed by a condensed description of analog computer problem-solving techniques. Don't expect that the next few pages will turn you into a math wizard; they won't. You may, however, come away with a better feel for *derivatives* and *integrals*, and you'll get some insight into the kinds of problems an analog computer is intended to solve.

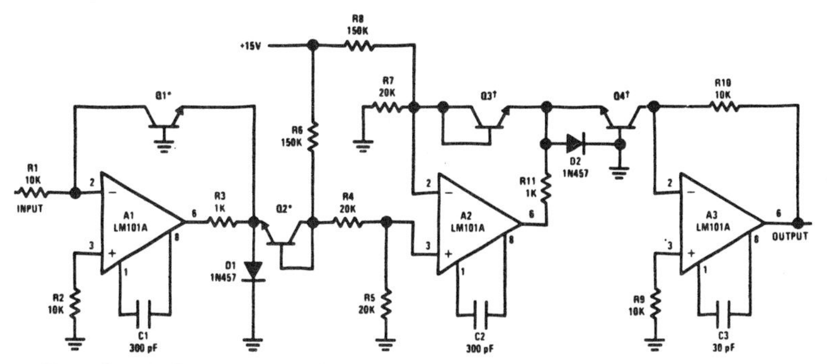

Fig. 9-11. Selection of different values for R4 and R5 produces different roots of the input volage. With R4 and R5 equal, the output is the square root of the input.

EQUATIONS AND VARIABLES

In an algebraic equation, we attempt to represent one quantity in terms of another, usually because the second quantity is easier to measure. The first quantity is called the *dependent variable*; the second, the *independent variable*. For example, Ohm's law states that the relationship between two quantities, voltage and current, that can be observed when a potential difference is applied across a resistance. If we know the current and are interested in finding the voltage, we can write Ohm's law as

$$E = R\,(I) \tag{7}$$

In this case, current I is the independent variable, voltage E is the dependent variable, and resistance R is a constant multiplier. We could just as well be interested in finding the current corresponding to a certain potential difference. In this case, we rewrite Ohm's law as

$$I = \frac{1}{R}\,(E) \tag{8}$$

Here, voltage E is the independent variable, current I is the dependent variable, and $1/R$ is a constant multiplier.

The relationship between dependent and independent variables can be more complicated than in the case of Ohm's law. Consider a cannonball fired upward at an angle θ, with a muzzle velocity V_o, as in Fig. 9-12. If we neglect air resistance and the curvature of the earth, we can come up with a not-too-complicated expression for the altitude y of the cannonball at any time after it is fired.

$$y = -\tfrac{1}{2}gt^2 + V_o\,(\sin\theta)\,t \tag{9}$$

The quantity g is the acceleration of gravity near the earth, approximately 980 cm/sec^2 (32 ft/sec^2), and t is time. We can also state an expression for height of the cannonball in terms of distance, x that it has traveled over the ground

$$y = x^2\left[\frac{-g}{2V_o^2\,(\cos^2\theta)}\right] + (\tan\theta) \tag{10}$$

In these expressions, y is the dependent variable and the independent variables are time t and horizontal distance x.

DIFFERENTIAL CALCULUS

Differential calculus is not the same thing as differential equations, but it is the first step in acquiring an understanding

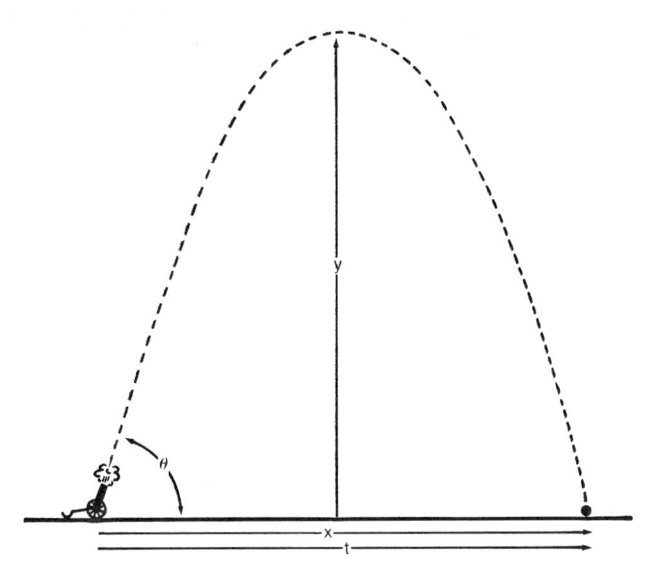

Fig. 9-12. In the equations in the text describing the flight of a cannonball, height **y** is the dependent variable, and distance **x** or time **t** is the independent variable.

of the study of differential equations. Differential calculus concerns itself with the *instantaneous rate of change* in a dependent variable with respect to an independent variable. Let's see what this means in the cases we have been discussing.

Slopes of Straight-Line Functions

Figure 9-13 is a graph of equation (7). We note that for any value of current *I* (the independent variable) such as I1 or I2,

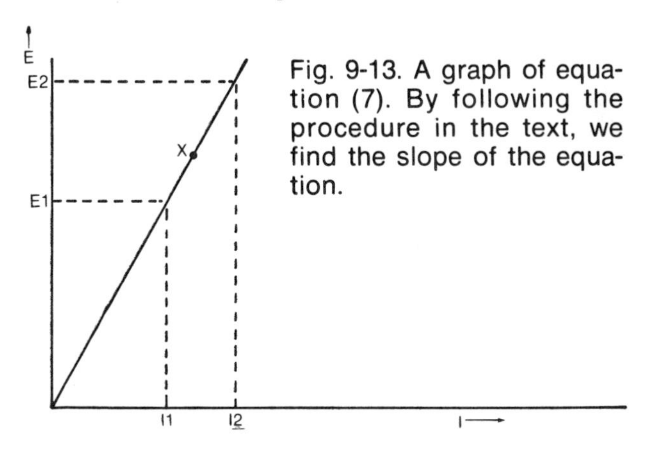

Fig. 9-13. A graph of equation (7). By following the procedure in the text, we find the slope of the equation.

there is a corresponding value of voltage E (the dependent variable). The values of E that correspond to I1 and I2 are E1 and E2.

Mathematicians define the *slope* of any curve (line on a graph) as the rate of change of the dependent variable with respect to the independent variable. Here, that would be the range of change of E with respect to I.

To find the slope at the point marked **X** on the curve in Fig. 9-13, we can take two points on the curve, one on either side of **X**, and determine the change in both E and I between them. The change in E, which we will call ΔE is E2 − E1. The change in I, or ΔI, is I2 − I1. The slope of the line is $\Delta E / \Delta I$.

We can obtain a unique expression for the slope of this kind of curve, if we express E2 and E1 in terms of I2 and I1, using equation (7)

$$E2 = R\,(I2) \qquad (11)$$

$$E1 = R\,(I1) \qquad (12)$$

Then ΔE, or $E2 - E1$, is $R\,(I2) - R\,(I1)$, or $R\,(I2 - I1)$, and

$$\frac{\Delta E}{\Delta I} = \frac{R(I2 - I1)}{(I2 - I1)} = R \qquad (13)$$

So the slope of E = R(I) is R.

Derivatives

The slope defined in the preceding paragraphs can only apply to linear equations—that is, equations that plot as straight lines. In differential calculus, mathematicians extend the usefulness of the idea of a slope so that it can apply to any sort of a curve. For example, take the curve representing the trajectory of the cannonball we looked at in Fig. 9-12. This has been redrawn with a few additions in Fig. 9-14. Suppose we wish to know the slope of the curve at point **X**. If we just took two random points on either side of **X**, say **A** and **B**, it is obvious that we wouldn't get anywhere near a true answer. So what must we do? Without going into the details of *how*, we can say that mathematicians select points a and b arbitrarily close to **X**, and then, speaking in a figurative mathematical sense, they allow those points to approach closer and closer to **X**. As the distance betwen points a and **X** and b and **X** approaches zero, the mathematician obtains a value for the slope that he calls the *derivative* of y with respect to x. He writes this dy/dx, to distinguish it from the cruder $\Delta y / \Delta x$ slope.

Since the derivative of a mathematical curve changes from point to point, it too can be represented by an equation

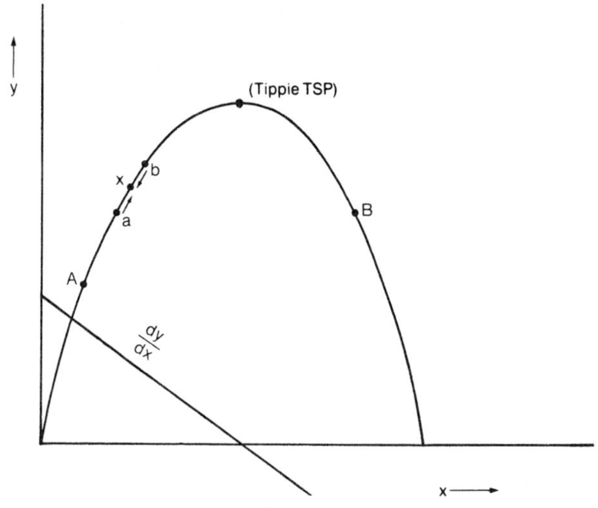

Fig. 9-14. The derivative of a continuously changing function is similar to the slope of a straight-line function. This is the cannonball trajectory again. The first derivative is plotted on the same scale as the height function. Note that the point at which it passses through zero corresponds to the apex of the trajectory.

and by a curve. Given the equation for the trajectory of the cannonball in equation (10), it is possible to obtain this expression for the derivative of that trajectory equation:

$$\frac{dy}{dx} = x\left[\frac{-g}{V_o^2(\cos^2\theta)}\right] + \tan\theta \qquad (14)$$

Why Derivatives?

Now you may ask what good all of this is. The specific utility of a derivative depends on the physical processes the equations are describing, but in this case, one thing we can use the derivative for is to tell us at what horizontal distance, x, the cannonball achieves its maximum altitude.

While the projectile is going up, both altitude and horizontal distance from the cannon are increasing so we know the sign of the derivative must be positive. After the cannonball passes the apex of its flight path and starts down, horizontal distance from the cannon is still increasing, (positive), but altitude is decreasing (negative). Thus, we know the derivative must have a negative sign. Then at the top of the trajectory, the sign of the derivative must be neither positive nor negative. The only way this can happen is if it is zero. Therefore, if we take equation (14), set it equal to zero, and solve for x, we will find the horizontal distance from the cannon at which the cannonball reaches its greatest height. We find this occurs when

$$x = \frac{V_o}{2g}(\sin 2\theta) \tag{15}$$

Second Derivatives, Third Derivatives, and So On

It isn't necessary to stop at one derivative. We can differentiate equation (14) to obtain the second derivative of equation (10).

$$\frac{d^2}{dx^2} = \frac{-g}{V_o^2(\cos^2 \Delta)} \tag{16}$$

The expression $\frac{d^2y}{dx^2}$

is read " the second derivative of y with respect to x." An infinite number of derivatives is theoretically possible. It happens that the third and all subsequent derivatives of equation (10) are zero, but there are other types of equations—for instance those in which the independent variable is a trigonometric or exponential function, for which an infinite number of derivatives may be found.

Physical Meanings of Derivatives

Derivatives of equations that have time for their independent variable are often themselves important physical quantities. Take equation (9), the expression for the displacement of the cannonball as a function of time. The derivative of this equation,

$$\frac{dy}{dt} = -gt + V_o(\sin \theta) \tag{17}$$

represents the *velocity* (in the y direction) of the cannonball. The second derivative of equation (9),

$$\frac{d^2y}{dt^2} = -g \tag{18}$$

represents the *acceleration* (also in the y, or vertical direction) of the cannonball. By the way, it is also correct to say that acceleration is the *first* derivative of velocity.

Derivatives with respect to time show up in electronics, also. For instance, the voltage across an inductor is equal to the inductance times the first derivative of current with respect to time.

$$e = L\left(\frac{di}{dt}\right) \qquad (19)$$

It happens that the derivative of *sin wt* with respect to *t* is *w cos wt*, so if the current flowing through an inductor of 10H is 0.12 *sin* 10*t*, or 120 mA at a frequency of 10 radians/sec, the voltage appearing across that inductor will be 12 *cos* 10*t*, that is, 12V at a frequency of 10 radians/sec.

INTEGRAL CALCULUS

You can think of integration as the opposite of differentiation, but that leaves a lot unsaid. Integration is basically a summing process. In algebra class, you may have encountered an expression like this:

$$y = \sum_{n=1}^{10} \frac{1}{n} \qquad (20)$$

This is a shorthand way of saying that

$$y = 1/1 + 1/2 + 1/3 + 1/4 + 1/5 + 1/6 + 1/7 + 1/8 + 1/9 + 1/10 \qquad (21)$$

Integration is another form of summation, but instead of summing quantities of a measurable size like those on the right side of the equal sign in equation (21), it sums an infinite number of infinitely small quantities. That probably sounds like nonsense, so let's look at a practical example to see if we can't make the idea a little more clear.

Hydrostatic Force Behind Dam

Figure 9-15 represents a dam in a river. The water is backed up behind the dam to a depth of *H* as shown. In our reference system, with $h = 0$ denoting the surface of the water, we can represent any arbitrary depth below the surface by a small letter *h*. The width of the dam from side to side is *W*.

Suppose we want to know the force that the water behind the dam exerts on the dam. What do we know about this hydrostatic force? For one thing, we know that *pressure* increases with depth. Anybody who has ever dived to the bottom of a swimming pool and felt the pressure on his ears knows that. Actually, we know from measurement and from induction that pressure increases linearly with depth. If we talk about pressure in terms of square feet, the relationship is

$$p = 62.4\,(h)$$

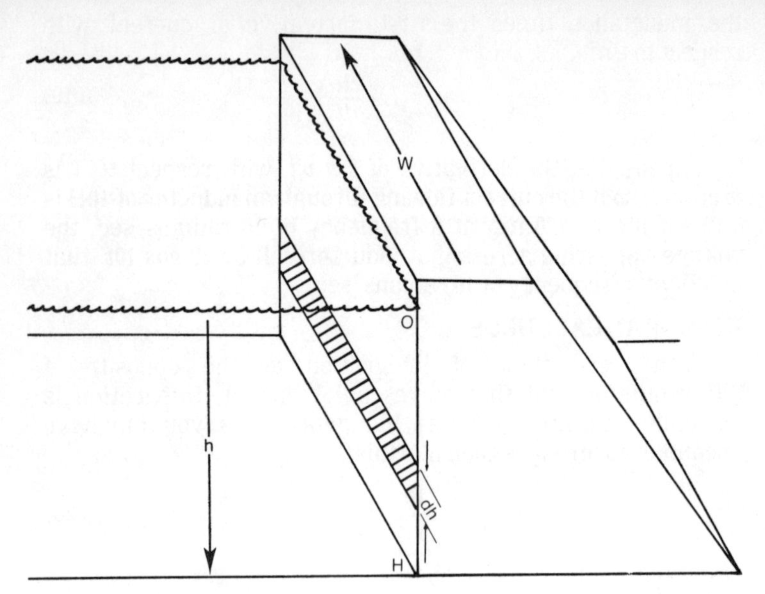

Fig. 9-15. This dam in a river problem helps to illustrate the ideas behind integral calculus. The problem is to find the force of the water backed up behind the dam.

where p is pressure and h is depth in feet. This tells us, for instance, that the pressure 10 feet below the surface is 624 lb/ft^2.

Pressure, though, is not the same thing as force. Pressure represents the force felt over some unit area, so to find the actual force, we must multiply pressure times area. Can we just multiply the area of the dam face times the pressure to find the force? The answer is no, since the pressure on the dam face isn't uniform, but varies with depth.

What about the pressure across a small area of the dam face, though—an area such as the shaded portion of height *dh* in the drawing? If *dh* corresponds to a foot of depth , say the foot from 9 to 10 ft, the pressure will vary from 562 lb/ft^2 at the top of *dh* to 624 lb/ft^2 at the bottom of the strip. If we narrow *dh* to only 6 in., say the 6 in. from 9.5 to 10 ft, the pressure will vary from about 593 lb/ft^2 to 624 lb/ft^2. As *dh* shrinks, the pressure difference across it shrinks also. If *dh* were infinitely thin, the pressure at its top and bottom would be the same.

To find the force of the water against the dam, then, we can divide the face of the dam up into a number of strips, each *W* ft wide and *dh* ft deep. If we make *dh* equal to 1 ft, there will be *H* strips, since we have said that the water is *H* ft deep. If we make *dh* equal to half a ft, there will be twice as many strips. If dh is infinitesimally small, there will be an infinite number of strips.

Given this information, along with the methods of integral calculus, we know enough to evaluate the force on the dam. The force the water exerts against any strip of depth dh is equal to the pressure at that depth times the area of the strip. Pressure, from the relationship given above, is $62.4\ (h)$. The area of any strip is just $W \times dh$. Thus, F_{strip}, the force against any strip is

$$F_{strip} = 62.4\ (h) \times W \times dh \tag{22}$$

Before we go to the calculus, suppose we make an approximation of the total force against the dam using ordinary summation. If we assume dh is 1 ft, and there are H strips, then the total force against the dam is the sum of the forces against those strips.

$$F = \sum_{h=1}^{H} 62.4\,(h) \times W \times 1 \tag{23}$$

If we make dh equal to 0.5 ft, there will be $2H$ strips, and

$$F = \sum_{h=1}^{2h} 62.4 \left(\frac{h}{2}\right) W \times 0.5 \tag{24}$$

Equations (23) and (24) will give only approximate answers, but (24) should be more accurate than (23) because it uses more and narrower strips. Let's assume some values for W and H and see what answers these two equations give us. Let's say that the water is 12 ft deep, or $H = 12$, and let's say further that the dam is 20 ft wide, or $W = 20$. Then, solving equation (23), we find

$$F = 62.4\ (78) \times 20 \times 1 = 97{,}200 \text{ lb}$$

Solving equation (24), we get

$$F = 62.4\ (150) \times 20 \times 0.5 = 93,600 \text{ lb}$$

Integral calculus is a method of carrying the procedure of equations (23) and (24) to its ultimate conclusion. The dh dimension of a strip is allowed to approach zero and the number of strips is allowed to approach infinity. The result is a precise answer for the force of the water against the dam. Replacing the summation sign in equation (23) with the sign for a definite integral, we have

$$F = \int_0^H 62.4\,(h) \times W \times dh \qquad (25)$$

Using the methods of integral calculus, we can solve this equation to obtain

$$F = 62.4 \left(\frac{H^2}{2}\right) \times W \qquad (26)$$

If we substitute the values of W and H used above, we get

$$F = 62.4\,(72) \times 20 = 89,856 \text{ lb}$$

This is the exact value of the force of the water acting against the dam.

Physical Meaning of Integrals

Integrals with time as the independent variable are very important in the physical world. If velocity is the first derivative of displacement with respect to time, and acceleration the first derivative of velocity with respect to time, then velocity must be the integral of acceleration and displacement must be the integral of velocity. There are similar processes that take place over periods of time, like chemical reactions and biological processes, for which integrals have important physical meaning.

In electronics, the integral of current with respect to time, $i\,dt$, represents charge. Thus, the voltage across a capacitor is given by

$$e = \frac{1}{C}\, i\,(dt) \qquad (27)$$

DIFFERENTIAL EQUATIONS

The study of differential equations is an extension of integral calculus. It is one of the most important keys to understanding any kind of science.

In utilitarian terms, analysis by means of differential equations starts with an observation about something. For instance, Newton, in his second law of motion, hypothesized that if a *force* acts on a body, the *rate of change* of the *momentum* of that body with respect to time is proportional to the force and codirectional with it. Linear momentum is equal to mass times velocity, so the rate of change of momentum with respect to time is $d\,(mv)/dt$. [We use the expression mv to represent momentum with the understanding that the v term includes both a magnitude (so many feet per second, miles per hour, etc.) and a direction (up, down, right, left, at an angle of 22° with the horizontal, etc.) (That is to say, we understand that v and mv and F are vector quantities.) Reduced to a mathematical expression, Newton's second law of motion is

$$F = \frac{d(mv)}{dt} \tag{28}$$

In most cases, the mass of an object remains constant. (The most important exception to this is the case where velocity approaches the speed of light), so equation (28) reduces in most cases to

$$F = m\left(\frac{dv}{dt}\right) \tag{29}$$

or, since dv/dt is acceleration, to just $F = m\,(a)$.

Now that's a beautiful insight to have running through our minds as we sit quietly watching apples fall, but how can we find out if it is true? We can't measure acceleration directly, nor can we measure velocity if it is constantly changing. The only thing we can measure easily is distance and time. What we need is a way to change equation (29) to an equation that deals with distance and time. The study of differential equations is the study of how to get from an equation involving differentials or integrals to one involving simple functions of variables.

In the case of equation (29), we can rewrite it as

$$\frac{dv}{dt} = \frac{F}{m} \tag{30}$$

As we have already noted, the integral of acceleration, dv/dt, is velocity, v, so we can use the methods of integral calculus to integrate both sides of equation (30), obtaining

$$v = \frac{F}{m}(t) + V_0 \tag{31}$$

Where V_0 is the initial velocity of whatever body it is we are studying.

Next, we need to go from this velocity equation to one involving displacements. We recall that velocity is the derivative of displacement, which we will call y, so that $v = dv/dt$. Then, rewriting equation (31),

$$\frac{dy}{dt} = \frac{F}{m}(t) + V_0 \tag{32}$$

Integrating both sides of this equation, using the rules of integral calculus, gives us

$$y = \frac{F}{2m}(t^2) + V_0(t) + y_0 \qquad (33)$$

Now we are in a position to check Sir Isaac's hypothesis. It is relatively easy to construct an experiment in which a constant force is applied to a given mass and to measure the displacement of that mass over several periods of time. With enough measurements, we can verify the theory and come up with a system of units for force and mass as well.

Analog Computers and Differential Equations

Those who have studied calculus find that the method of going from equation (28) to equation (33) is direct and fairly easy to understand. Not all differential equations, however, are analyzed so easily. There are general procedures for attacking different classes of differential equations, but using them requires practice and sometimes a good deal of art.

Many, many physical systems, mechanical, electrical, chemical, etc., can be described by sets of differential equations. When these differential equations get sufficiently complicated, there are two effective methods for dealing with them. One is a paper and pencil method, using the Laplace transform, which we will not discuss here. The other method involves the use of the electronic analog computer to simulate directly the system described by the differential equations.

The Op-Amp as Integrator

In these cases, the key analog computer component is the operational amplifier connected as an integrator—that is, as a device with an output voltage that is the mathematical integral of its input voltage. The integrator is essentially an op-amp with a resistance in series with its inverting input and with a capacitor connected as the feedback impedance from its output to its inverting input (Fig. 9-16).

To see how this works, let's go back to the procedure we used to analyze the op-amp with input and feedback impedances in Chapter 1. Recall that we said that because of the high input impedance, input current *I*1 had to be equal to feedback current, I2. We also said that because of the high gain of the op-amp, the voltage at node *A* in the figure was essentially zero as long as the output voltage was below saturation.

Under these circumstances, the input current *into A* (*I*1) is

$$I1 = \frac{-V_{in}}{R} \qquad (34)$$

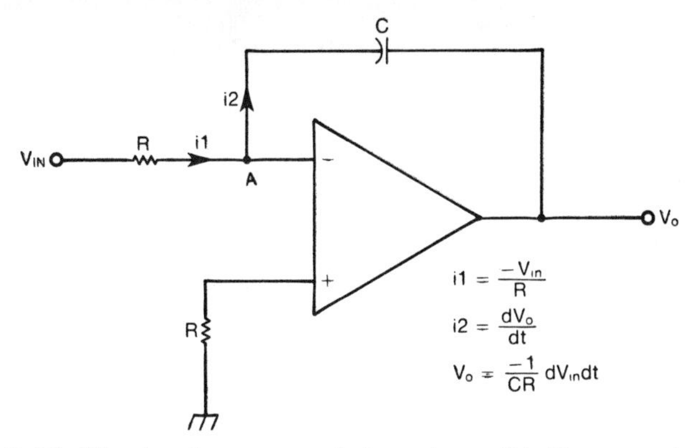

Fig. 9-16. The basic op-amp integrator with its associated equations.

Then, integrating equation (27), we can say that the feedback current, *I2*, *out of* node *A* is

$$I2 = C\left(\frac{dV_o}{dt}\right) \tag{35}$$

Setting the two currents equal,

$$\frac{-V_{in}}{R} = C\left(\frac{dV_o}{dt}\right) \tag{36}$$

Finally, by rearranging the terms and integrating both sides,

$$V_o = \frac{-1}{RC}\int V_{in}dt \tag{37}$$

That is, the output voltage is the negative of the integral of the input voltage, times the reciprocal of the feedback capacitance in fardds, times the input resistance in ohms.

Programing Symbols

The analog computer programer uses three circuit symbols based on the triangle we have already employed to represent the basic operational amplifier. A summing amplifier, or summer, is represented as in Fig. 9-17A. The numbers adjacent to the inputs indicate the gain of the summer at that junction. The integrator is represented by an op-amp triangle with a rectangular extension on the input side, as in Fig. 9-17B. It may have a single input, or it may have several summing inputs. The *initial condition* input, at the top of the rectangle, allows the integrator output to be set to some nonzero value before the integration process begins.

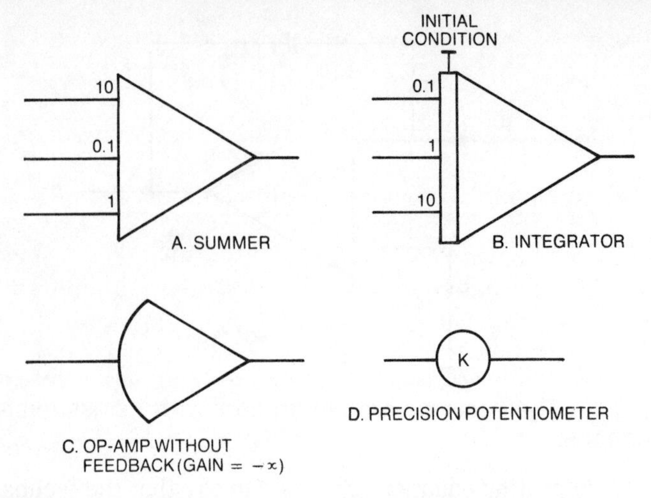

Fig. 9-17. The four symbols used by the analog computer programer.

Sometimes, an analog computer op-amp is used as a comparator. In this case, the op-amp without feedback is drawn with a curved back, as in Fig. 9-17C, and the external resistors are drawn in as necessary. The fourth symbol used on analog computer programing diagrams is the circle, Fig. 9-17D, representing a precision potentiometer. Most commercial analog computers use 10-turn pots that allow resistance to be set to 3-digit precision, anywhere from 0.001 to 0.999. By combining pot attenuation with decade amplifier gains, the programer can precisely set system gains from 0.0001 to 999. Summers can be cascaded for even greater gain, but this usually introduces problems.

Using the Analog Computer

Let's take a very basic example to see how a differential equation is simulated on an analog computer. Suppose we have a hypothetical system consisting of a weight and a spring. The force that the spring exerts on the weight f is proportional to a constant k the spring exerts on the weight y, or

$$f = k(y) \qquad (38)$$

The minus sign indicates that the force is in the opposite direction from the displacement. Equation (3) tells us that the acceleration (dv/dt, or d^2y/dt^2) of the weight is proportional to the net force on the weight divided by the weight's mass, so by combining equations (30) and (38),

$$\frac{d^2y}{dt^2} = \frac{-k}{m}(y) \qquad (39)$$

This is the differential equation that describes the motion of the weight on the spring.

As soon as we decide on one other factor, we can proceed to set up an analog computer program to simulate equation (39). That factor is the initial displacement of the weight. Obviously, if nothing disturbs the spring and weight, they will both sit there without moving. With no initial input, the solution to equation (39) is just $y = 0$, but that's trivial. Let us say that we initally displace the weight some y_0 centimeters from its rest position. The initial conditions then are

$$y_{(t=0)} = y_0 \qquad (40)$$

That is, y, at time t equal to zero, is y_0.

There are four steps in preparing the analog computer diagram that represents equations (39) and (40). These are shown in Fig. 9-18. In 9-18A, we start with a single integrator. We call the *input* to this integrator d^2y/dt^2. The *output* of this integrator is then $-dy/dt$.

We can't do anything with just $-dy/dt$, so we add another integrator, as in Fig. 9-18B. With another integration and another sign change, we now have a voltage that simulates y itself. Thus we have two leads, the one on the left representing the variable on the left of equation (39), the other representing the variable (but not the constant multiplier) on the right of that equation. In Fig. 9-18C, we have added a summer and a potentiometer to the two integrators. The attenuation of the potentiometer and the gain of the summer are selected to make the output of the summer equal to $- k/m\,)y)$. Note that the summer introduced another sign change. Every amplifier in an analog computer simulation, regardless of whether it is used as a summer or as an integrator, will introduce a sign change.

Now that we have a voltage representing $- k/m\,)y)$, all we need do to make the computer diagram the same as the equation is to connect the output of the summer to the input of the first integrator. This, after all, is what the equal sign in the equation means. We must also connect a reference voltage through a potentiometer to the initial condition input of the second integrator. We adjust the potentiometer until the output voltage of that integrator is equal to y_0. Both of these steps have been done in Fig. 9-18D.

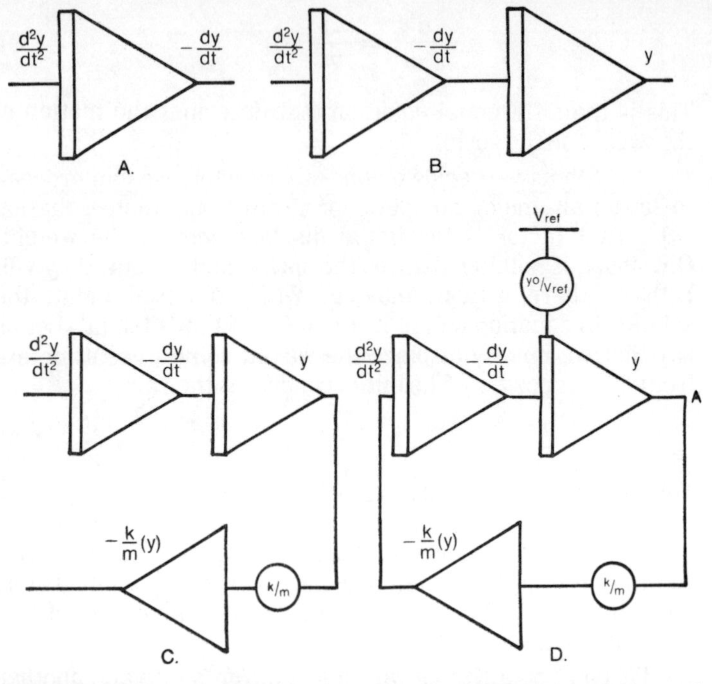

Fig. 9-18. Four steps in the analog computer simulation of equations (39) and (40).

If we interconnect the amplifier of a real analog computer to match the diagram of Fig. 9-18D, we can connect an oscilloscope or x-y recorder to point A on the circuit to observe what happens to y over a period of time. In this case, if we connect a scope and then apply power to the amplifiers (simultaneously removing the reference voltage supplying y_o), we will observe that the voltage representing y follows a cosine function. The amplitude of the voltage will be $\pm y_o$, and the frequency will be equal to the square root of k/m.

As a numerical example, suppose the spring constant were 90 dynes/cm (the dyne is a unit of force in the metric system), and the mass of the weight were 10 grams. Then, regardless of the initial displacement, the frequency of oscillation of the mass and spring would be the square root of 9, or 3 radians/sec (about 0.48 Hz). If the initial displacement of the weight were 2 cm, and if this were represented by a potential of 2V at the output of the second integrator, the amplitude of the sine wave would be 2V. In mathematical terms, the solution to

$$\frac{d^2y}{dt^2} = -9; \; y_{(t=0)} = 2 \tag{41}$$

would be

$$y = 2 \cos 3t \tag{42}$$

Further Considerations—Damping

It has probably occurred to you that in real life, if you take a weight suspended from a spring and pull and release the weight, it does not forever oscillate at a constant amplitude; rather, the oscillations gradually die away. This illustrates an important limitation to be considered any time you encounter a differential equation that supposedly simulates a real-life system: the differential equation may be oversimplified.

To approach the behavior of a spring and mass system more exactly, it is necessary to introduce a new term in the differential equation. In this case, we decide that the force on the weight is proportional not only to displacement, but to the velocity at which the weight is moving. This is presumed to account for air resistance, and for entropic losses in the spring. So, instead of equation (38), we have

$$F = -k1\left(\frac{dy}{dt}\right) - k2\,(y) \tag{43}$$

Again using equation (30), we arrive at the differential equation

$$\frac{d^2y}{dt^2} = \frac{-k1}{m}\left(\frac{dy}{dt}\right) - \frac{k2}{m}(y) \tag{44}$$

We proceed as before to draw an analog computer diagram that corresponds to this equation, and we arrive at the diagram in Fig. 9-19A. This is similar to the one in Fig. 9-18D, except that we have added a potentiometer so that the input to the first integrator is the sum of two functions, one of y itself and the other of the first derivative of y.

This time, an oscilloscope connected at point A would show a voltage waveform like one of in Fig. 9-19B.

$$y = \frac{k1}{2m^t}(\cos \omega t) \tag{45}$$

where

$$\omega = \frac{1}{2m}\sqrt{k2\,(\mathrm{m}) - \frac{k1^2}{4}} \tag{46}$$

Mathematicians and scientists frequently encounter differential equations like equation (44). In fact, this is so common, there is a widely recognized general form of this

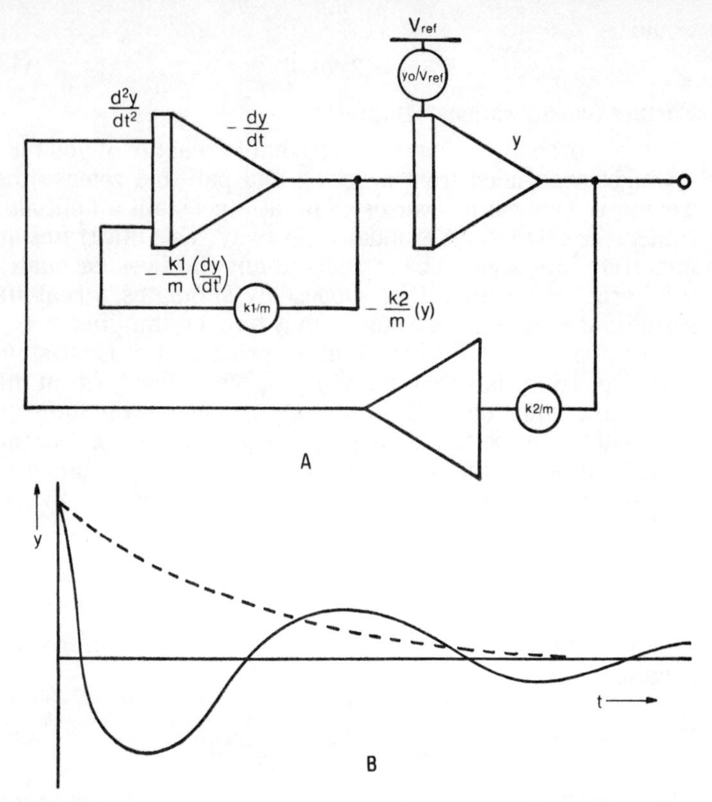

Fig. 9-19. The weight and spring system with damping. The output signal from the analog computer will take on one of the forms shown in B.

differential equation and some important characteristics are defined by it. The general form of equation (44) is

$$\frac{d^2y}{dt^2} + 2\zeta\omega_n \left(\frac{dy}{dt}\right) \omega_n^2(y) = 0 \tag{47}$$

The quantity ζ is called the *damping ratio*. It represents the rate at which the system oscillations will die away. The quantity ω_n is called the *undamped natural frequency*, the frequency (in radians/sec) at which the system would oscillate were it not for the damping.

The general solution to equation (47) is

$$y = y_o \, e^{-qt} (\cos \omega t)$$

where y_o is the initial value of y, and $q = \zeta\omega_n$. The quantity ω is the actual frequency of oscillation. It can be calculated from

$$\omega = \omega_n\sqrt{1 - \zeta^2} \quad (49)$$

Differential equations of the form of equation (47) are common in electronics as well as in mechanics. Figure 9-20A shows a circuit in which a capacitor of C farads is in series with a resistor of R ohms. A switch allows an initial condition current of V/r amperes to be established in an inductor of L henrys. At time t equals zero, the switch is thrown and the coil is placed in series with the resistor and the capacitor. If we call the current in the loop, i, we can set the sum of the voltage drops around the loop equal to zero.

$$L\left(\frac{di}{dt}\right) + R\,(i) + \frac{1}{C}\int i\,dt = 0 \quad (50)$$

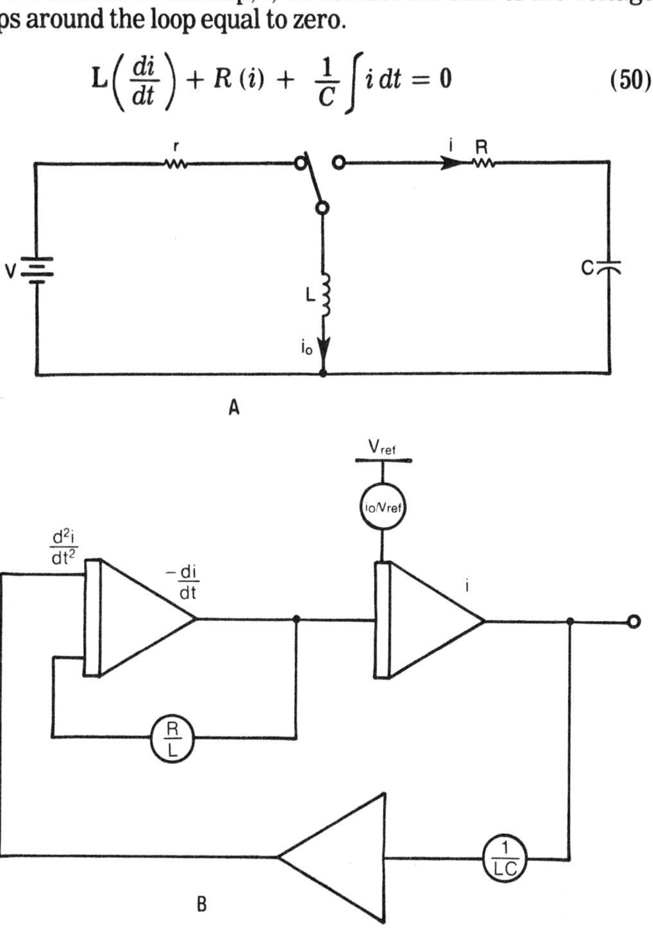

Fig. 9-20. An electronic circuit and its analog computer simulation. Once the computer simulation is set up, the effect of changing various components can be observed merely by changing pot settings.

If we integrate once and divide through by L, we obtain

$$\frac{d^2i}{dt^2} + \frac{R}{L}\left(\frac{di}{dt}\right) + \frac{1}{LC}(i) = 0 \qquad (51)$$

which is in the form of equation (47). The analog computer diagram of equation (51) in Fig. 9-20B is similar to the computer diagram in Fig. 9-19a.

Viewing the Output

Generally, we can display analog computer results on any of three kinds of devices. The most economical approach is the strip chart recorder, in which an electric motor drives a strip of paper at a preset rate past a pen arm. The pen is usually connected to a galvanometer movement which is driven by one of the amplifiers in the computer simulation. Another display device is the familiar oscilloscope. The computer output is connected to the scope's vertical input and a sweep voltage is used to drive the spot horizontally. We can save the output display either on the scope itself, if we have an expensive, variable-persistence scope, or we can simply take a time exposure photograph of the scope display. The x-y plotter is a device that produces a display similar to that of the oscilloscope. It consists of a table that holds a sheet of graph paper and a movable pen controlled by a pair of servomechanisms. As in the case of the scope, the analog computer output is usually connected to the y servo input of the recorder, while a sweep voltage drives the x servos.

Generating Ramp Functions

An oscilloscope sweep function is a matter of some practical importance. Most oscilloscopes do have their own sweep generators, but frequently they sweep at too fast a rate for analog displays. Many x-y recorders have no sweep generators of their own at all. This does not present a problem, however, since the analog computer can itself generate the needed sweep signal by means of a single integrator. Figure 9-21A shows an integrator with an input from the computer's negative reference-voltage supply. This is applied through a potentiometer. Figure 9-21B is a graph of the integrator's output. If we call the potentiometer setting $1/k$ the integrator output will increase linearly and will reach the reference-voltage level (and will saturate) k seconds after power is applied to the system. In this way, we obtain any sweep rate we desire.

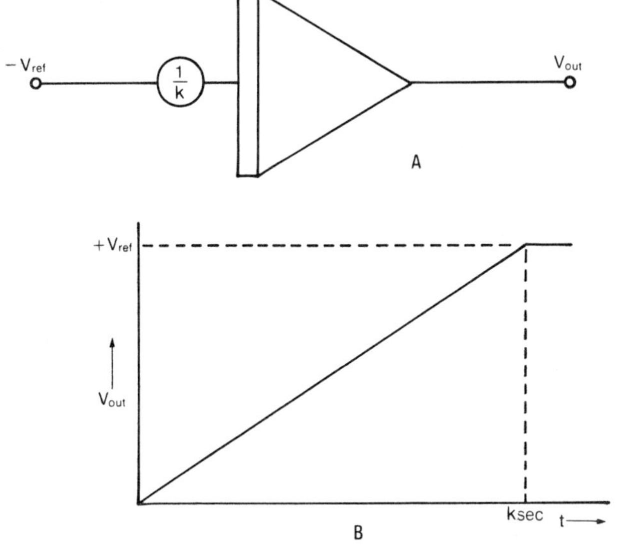

Fig. 9-21. Using an integrator to generate a ramp function. This is a useful way of obtaining a sweep signal with which to drive the **x** input of an output display.

Time Scaling

Often, the real advantage of an analog computer arises from its ability to change the time scale within which a process takes place. Something that happens too rapidly to observe can be slowed down, or something that takes place over a long period of time can be speeded up.

As an example, take the electronic circuit in Fig. 9-20A. Suppose we assign the following values to the circuit

$$V = 100\text{V} \qquad r = 20\Omega$$
$$C = 1\mu\text{F} \qquad R = 16\Omega$$
$$L = 80\,\text{mH}$$

Note that this produces an initial condition of a 5A current flowing in the coil at time t equals 0. With these circuit values, the differential equation is

$$\frac{d^2i}{dt^2} + 200\left(\frac{di}{dt}\right) + 1.25 \times 10^7(i) = 0 \qquad (52)$$

Relating equation (52) to equation (47), we find that the undamped natural frequency of the current oscillations will be about 3640 radians/sec (580 Hz) and that the exponential term in equation (48) will be e^{-100t}. Both of these are far too fast for the analog computer, which begins to lose accuracy when frequencies exceed about 1 Hz, and for most mechanical display devices.

Fortunately, it is easy to change the time scale of a differential equation without affecting any of the amplitude factors. The rule is as follows:

To slow a differential equation by a factor A, multiply each second derivative term by A^2, and each nth derivative term by A^n. To speed up a differential equation by a factor A, divide each nth derivative term by A^n.

In our example, if we slow the equation by a factor of 1000, the undamped natural frequency will be a little more than 0.5 Hz, and it will take fully 10 seconds for the waveform envelope to decrease to 37% of its starting value. This seems like a reasonable choice of time scale, so we begin by multiplying the second derivative term in equation (52) by 1000^2 and the first derivative term by 1000. The i term, the *zero*th derivative, if you will, is multiplied by 1000^0, or 1.

$$10^6\left(\frac{d^2i}{dt^2}\right) + 2 \times 10^5\left(\frac{di}{dt}\right) + 1.25 \times 10^7(i) = 0 \qquad (53)$$

Now we divide through by 10^6

$$\frac{d^2i}{dt^2i} + 0.2\left(\frac{di}{dt}\right) + 12.5\,(1) = 0 \qquad (54)$$

Here, as we wanted, the undamped natural frequency is 3.64 radians/sec and the time constant is 0.1. Figure 9-22 is an analog-computer diagram of equation (54). All the amplitudes in the computer solution will correspond to amplitudes in equation, but each second of elapsed time in the computer solution will correspond to a millisecond of elapsed time in the real circuit described by equation (52).

Amplitude Scaling

Choosing an amplitude scale is as important as choosing a time scale in analog computer simulations. In the example we have been considering, the initial condition is $i_{(t=0)} = 5A$. Current i is represented by the output voltage of the second integrator in the analog circuit, and di/dt is represented by the output voltage of the first integrator. It is up to us to choose a factor relating current in the theoretical circuit to voltage in the analog circuit. There are two things to watch: We must not allow the introduction of errors or distortions either by allowing the signal voltage in some aplifiers to get so low that the signal-to-noise ratio becomes unmanageable or by allowing the signal voltage to get so large in any of the amplifiers as to cause saturation.

Most selections of amplitude scale factors is based on our knowledge of the real-world system simulated by the differential equation. In our example, we know that i never exceeds its initial condition value, but what of di/dt? It is possible to show that in a lightly damped second-order system, the maximum value of the first derivative is approximately equal to the maximum value of the function (in this case, i) times the natural frequency. Thus, for our circuit, the approximate maximum value of di/dt is 18.4.

Most modern solid-state analog computers use 10V reference supplies and $\pm$10V supplies for their amplifier V_{cc} requirements, so we want to avoid any values for the analogs of i and di/dt in excess of 10V. Ideally, we would like to make 1V in the analog circuit equal 1A in the real circuit, but this would cause the first integrator to go into saturation sometime during our run. Instead, we can settle on a scale factor of 1V = 2A. Once we have decided on this, all that is necessary is to set our initial conditions and inputs with this in mind. In Fig. 9-22, a potentiometer setting of 0.25 together with a reference

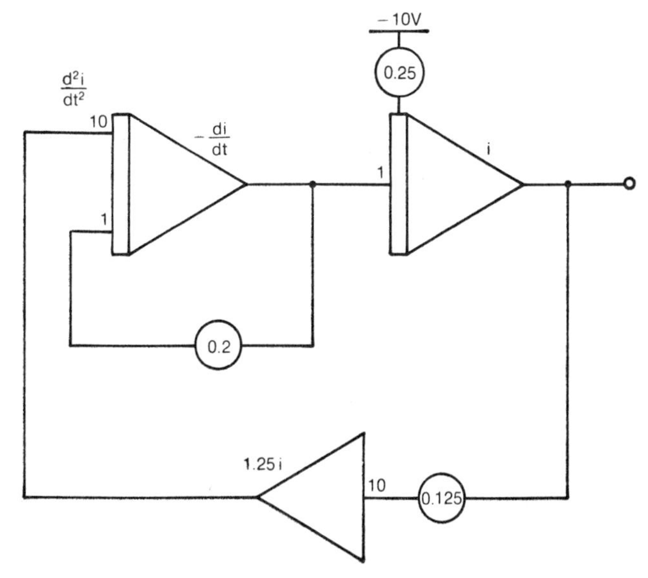

Fig. 9-22. An analog computer simulation of equation (52) with all time and amplitude scaling accomplished.

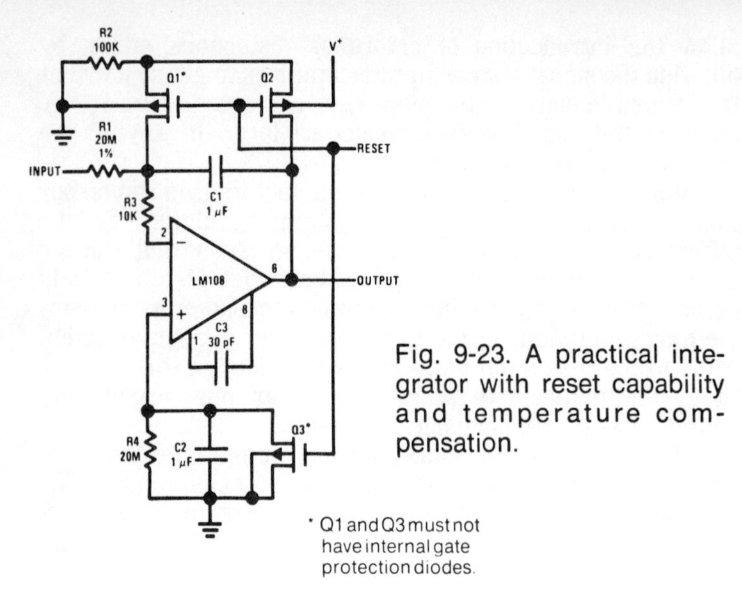

Fig. 9-23. A practical integrator with reset capability and temperature compensation.

* Q1 and Q3 must not have internal gate protection diodes.

voltage of 10V simulates the 5A initial condition in our example. When it comes time to interpret the data from the analog computer run, we will have to read each 1V graduation as 2A.

A Practical Integrator

A practical integrator circuit is shown in Fig. 9-23. The circuit uses a single National Semiconductor LM108 op-amp. It provides a solid-state reset capability and temperature compensation. The two *p*-channel MOSFET devices above the integrator are cut off, except when a negative reset signal is applied to their gates. In the off condition, there is essentially no voltage across Q1, so the leakage current of Q2 is absorbed entirely by R2. With the reset signal applied, both transistors conduct and C1 discharges, returning the integrator output to zero.

The portion of the circuit consisting of C2, R4, and Q3 provides temperature compensation if the characteristics of C2 and R4 are matched to those of C1 and R1. When the reset signal is applied to Q1 and Q2, it is also applied to Q3, shorting out the error voltage accumulated on C2.

The purpose of resistor R3 in the inverting input of the amplifier is to protect the input stage from damage that could occur if the amplifier were shut down with a charge on C1.

This is a numerically sequenced presentation of base diagrams and significant electrical characteristics of linear integrated circuits mentioned in the text. For detailed information on electrical characteristics of any particular linear IC, the reader should contact the manufacturer or his representative.

LH0021 POWER OP-AMP

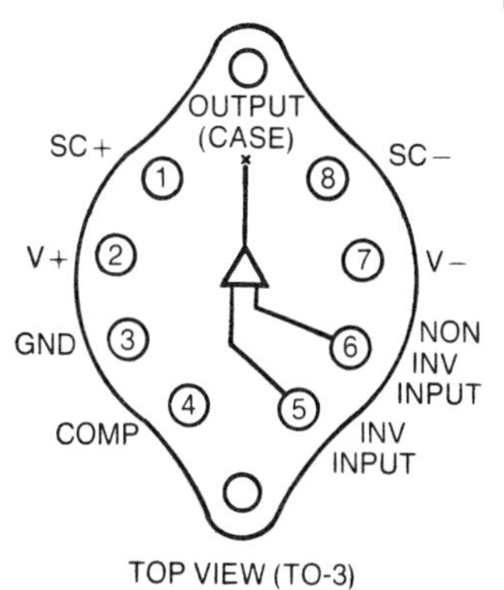

TOP VIEW (TO-3)

MANUFACTURER	NATIONAL SEMICONDUCTOR
SUPPLY VOLTAGE	±18V
SUPPLY CURRENT	4.0 mA MAX.
OFFSET VOLTAGE	6.0 mV MAX.
INPUT BIAS CURRENT	500 nA MAX.
VOLTAGE GAIN	100, 000 MIN.
CMRR	70 DB MIN.
BANDWIDTH	40 kHz

LH0041 POWER OP-AMP

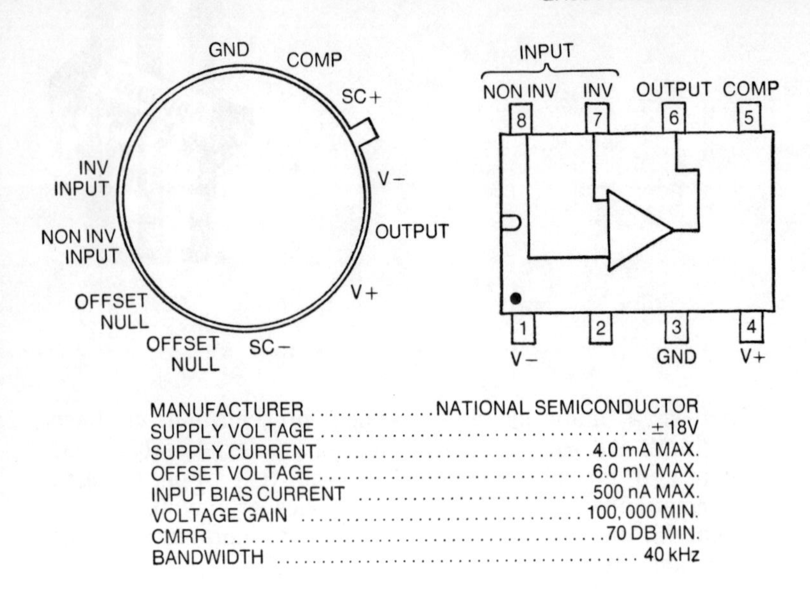

MANUFACTURER	NATIONAL SEMICONDUCTOR
SUPPLY VOLTAGE	±18V
SUPPLY CURRENT	4.0 mA MAX.
OFFSET VOLTAGE	6.0 mV MAX.
INPUT BIAS CURRENT	500 nA MAX.
VOLTAGE GAIN	100, 000 MIN.
CMRR	70 DB MIN.
BANDWIDTH	40 kHz

LH0042
LOW COST FET OP-AMP

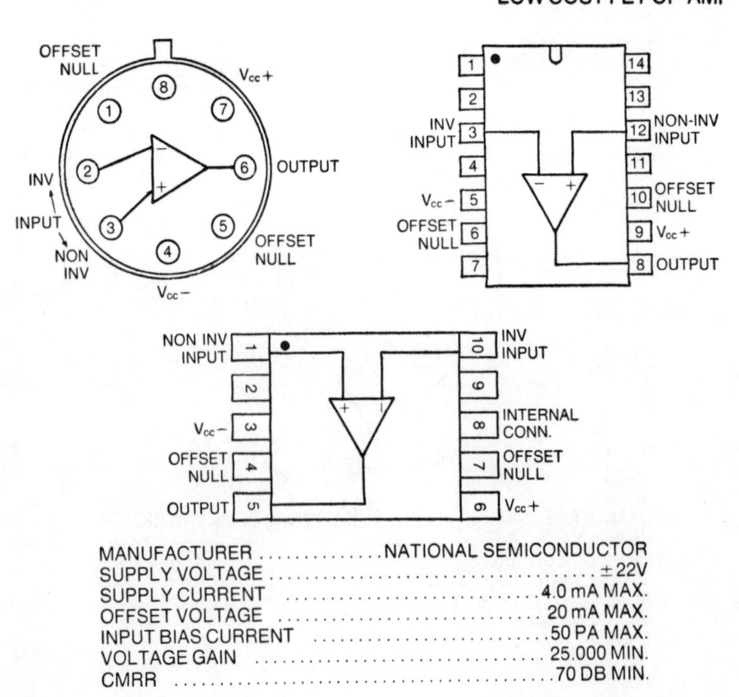

MANUFACTURER	NATIONAL SEMICONDUCTOR
SUPPLY VOLTAGE	±22V
SUPPLY CURRENT	4.0 mA MAX.
OFFSET VOLTAGE	20 mA MAX.
INPUT BIAS CURRENT	50 PA MAX.
VOLTAGE GAIN	25.000 MIN.
CMRR	70 DB MIN.

LH0062
HIGH SPEED FET OP-AMP

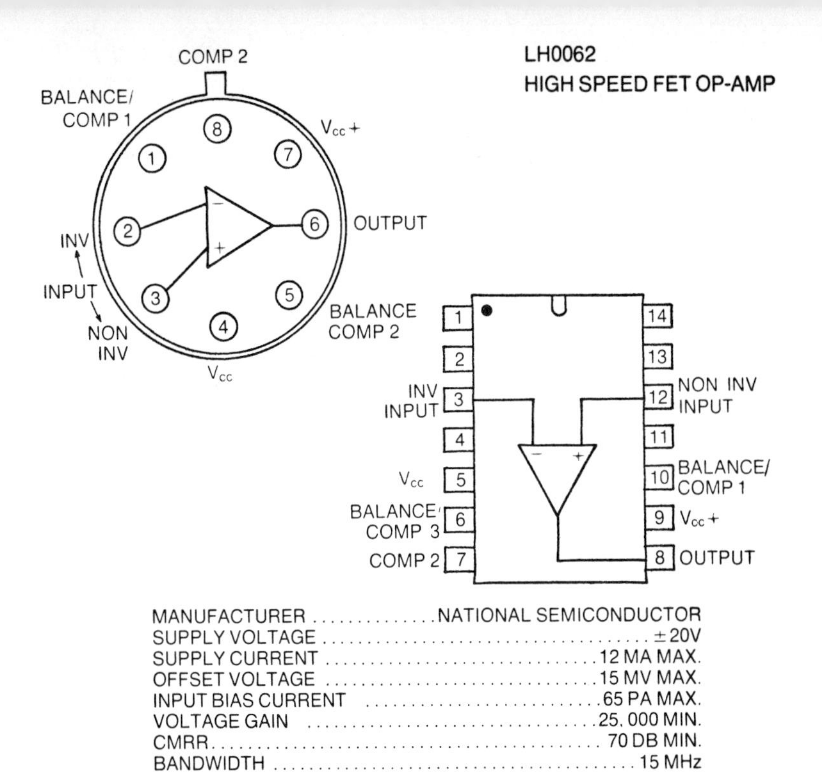

MANUFACTURER	NATIONAL SEMICONDUCTOR
SUPPLY VOLTAGE	±20V
SUPPLY CURRENT	12 MA MAX.
OFFSET VOLTAGE	15 MV MAX.
INPUT BIAS CURRENT	65 PA MAX.
VOLTAGE GAIN	25, 000 MIN.
CMRR	70 DB MIN.
BANDWIDTH	15 MHz

LH0063
BUFFER AMPLIFIER

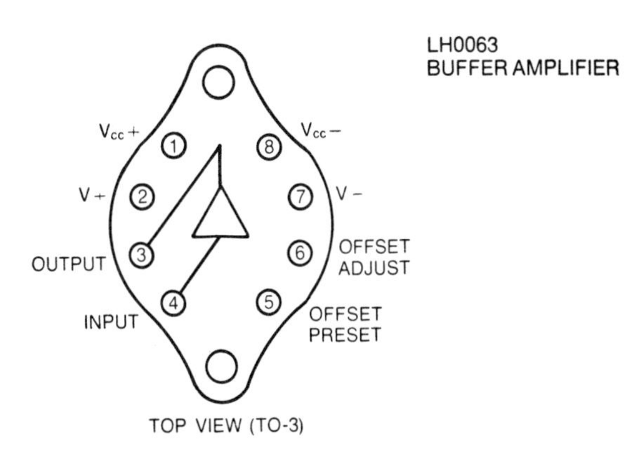

TOP VIEW (TO-3)

MANUFACTURER	NATIONAL SEMICONDUCTOR
SUPPLY VOLTAGE	±40V
SUPPLY CURRENT	24 mA MAX.
OFFSET VOLTAGE	20 mV MAX.
INPUT BIAS CURRENT	5 nA MAX.
VOLTAGE GAIN	0.98 TYP.
INPUT IMPEDANCE	10^{10} OHMS MIN.
OUTPUT IMPEDANCE	10 OHMS MAX.
SLEW RATE	1000 V/μs MIN.
BANDWIDTH	100 MHz

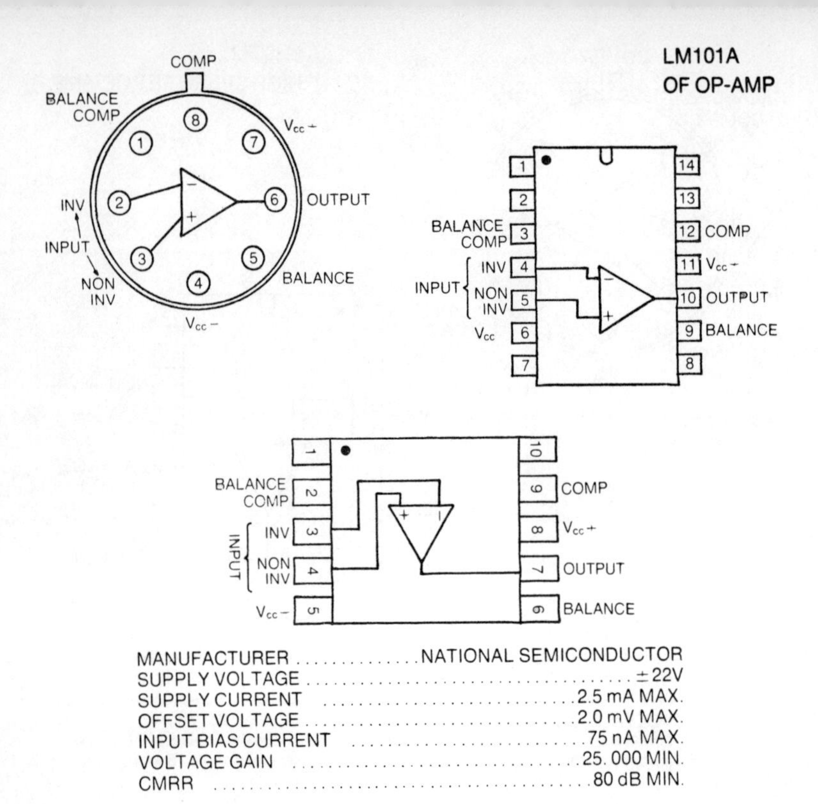

LM101A
OF OP-AMP

MANUFACTURER .... NATIONAL SEMICONDUCTOR
SUPPLY VOLTAGE .... ±22V
SUPPLY CURRENT .... 2.5 mA MAX.
OFFSET VOLTAGE .... 2.0 mV MAX.
INPUT BIAS CURRENT .... 75 nA MAX.
VOLTAGE GAIN .... 25. 000 MIN.
CMRR .... 80 dB MIN.

LM102
VOLTAGE FOLLOWER

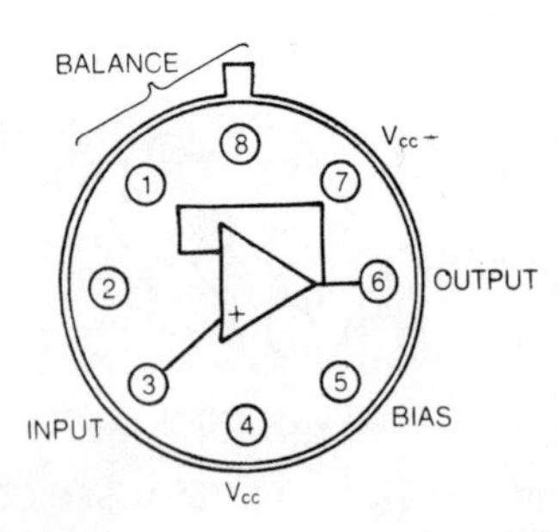

MANUFACTURER .... NATIONAL SEMICONDUCTOR
SUPPLY VOLTAGE .... ±18V
SUPPLY CURRENT .... 4.0 mA MAX.
OFFSET VOLTAGE .... 5.0 mV MAX.
INPUT BIAS CURRENT .... 110 nA MAX.
VOLTAGE GAIN .... 0.999 MIN
INPUT IMPEDANCE .... 10^{10} OHMS MIN.
OUTPUT IMPEDANCE .... 2.5 OHMS MAX.

LM104 NEGATIVE REGULATOR

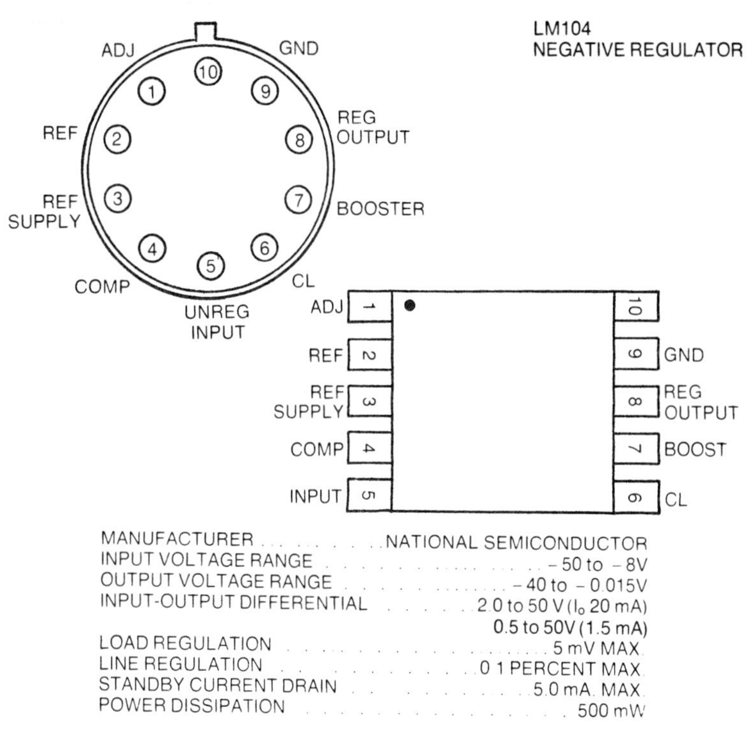

MANUFACTURER	NATIONAL SEMICONDUCTOR
INPUT VOLTAGE RANGE	−50 to −8V
OUTPUT VOLTAGE RANGE	−40 to −0.015V
INPUT-OUTPUT DIFFERENTIAL	2.0 to 50 V (I_o 20 mA)
	0.5 to 50V (1.5 mA)
LOAD REGULATION	5 mV MAX.
LINE REGULATION	0.1 PERCENT MAX.
STANDBY CURRENT DRAIN	5.0 mA. MAX.
POWER DISSIPATION	500 mW

LM105 POSITIVE REGULATOR

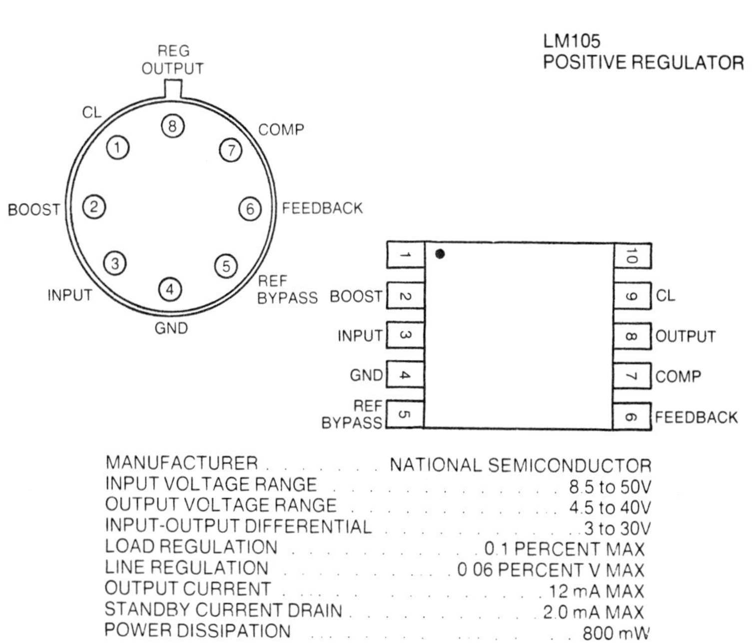

MANUFACTURER	NATIONAL SEMICONDUCTOR
INPUT VOLTAGE RANGE	8.5 to 50V
OUTPUT VOLTAGE RANGE	4.5 to 40V
INPUT-OUTPUT DIFFERENTIAL	3 to 30V
LOAD REGULATION	0.1 PERCENT MAX
LINE REGULATION	0.06 PERCENT V MAX
OUTPUT CURRENT	12 mA MAX
STANDBY CURRENT DRAIN	2.0 mA MAX
POWER DISSIPATION	800 mW

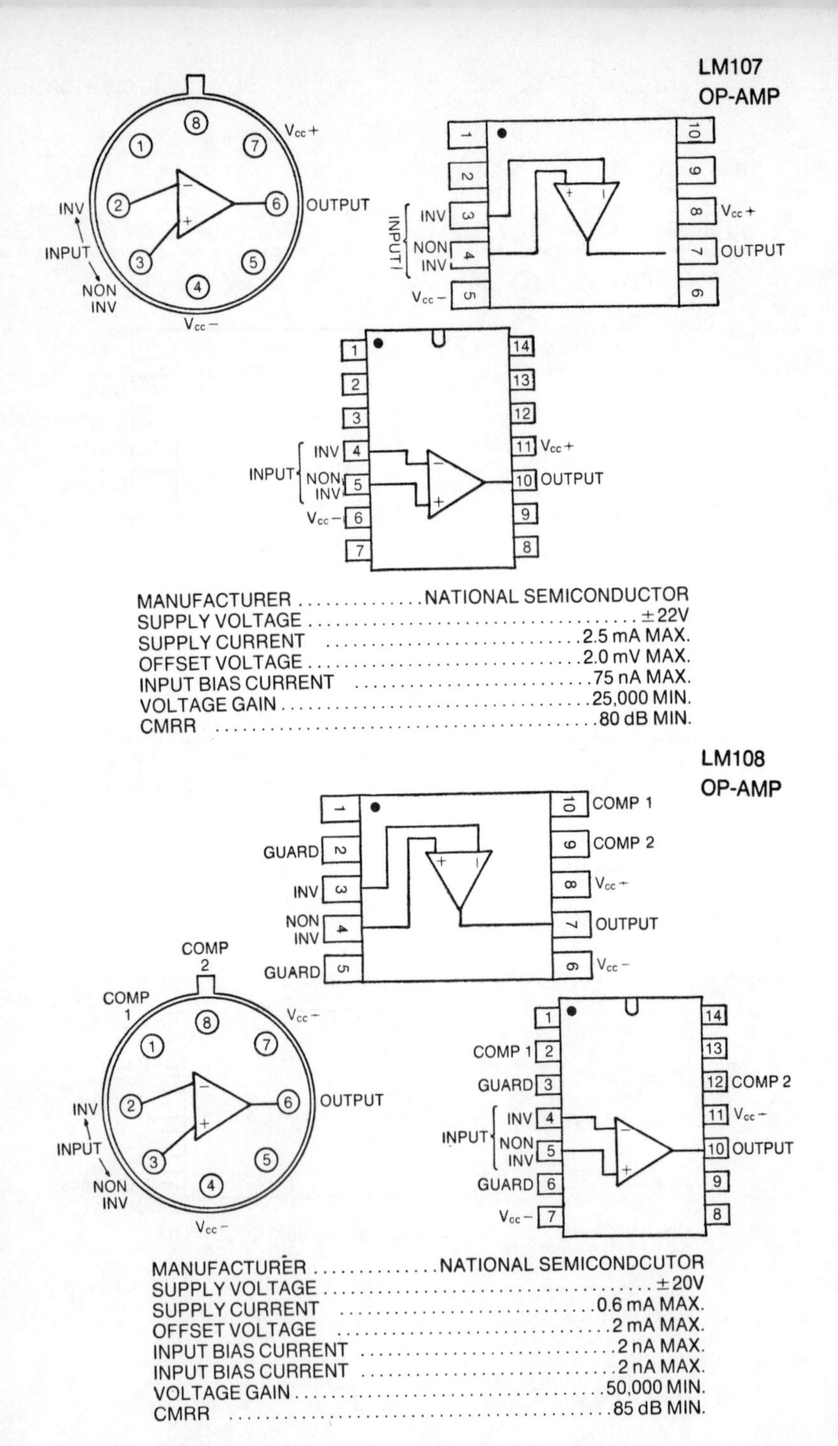

LM107
OP-AMP

MANUFACTURER .... NATIONAL SEMICONDUCTOR
SUPPLY VOLTAGE .... ±22V
SUPPLY CURRENT .... 2.5 mA MAX.
OFFSET VOLTAGE .... 2.0 mV MAX.
INPUT BIAS CURRENT .... 75 nA MAX.
VOLTAGE GAIN .... 25,000 MIN.
CMRR .... 80 dB MIN.

LM108
OP-AMP

MANUFACTURER .... NATIONAL SEMICONDCUTOR
SUPPLY VOLTAGE .... ±20V
SUPPLY CURRENT .... 0.6 mA MAX.
OFFSET VOLTAGE .... 2 mA MAX.
INPUT BIAS CURRENT .... 2 nA MAX.
INPUT BIAS CURRENT .... 2 nA MAX.
VOLTAGE GAIN .... 50,000 MIN.
CMRR .... 85 dB MIN.

LM109 VOLTAGE REGULATOR

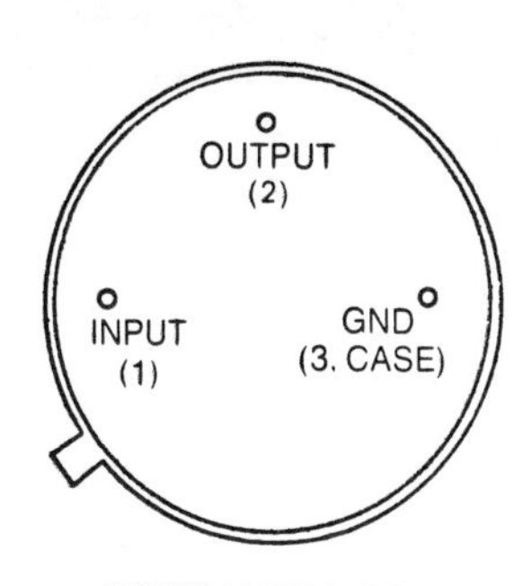

BOTTOM VIEW (TO-5)

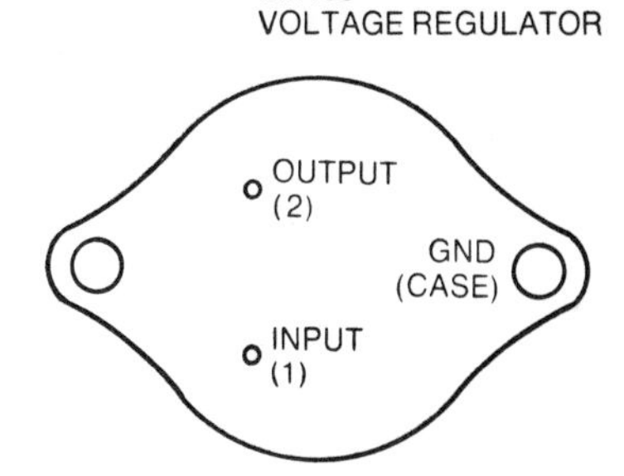

BOTTOM VIEW (TO-3)

MANUFACTURER .... NATIONAL SEMICONDUCTOR
INPUT VOLTAGE .... 35V MAX., 7V MIN.
OUTPUT VOLTAGE .... 5 V NOMINAL
LOAD REGULATION .... 50 mV MAX. (TO-5 PACKAGE)
100 mV MAX. (TO-3 PACKAGE)
LINE REGULATION .... 50 mV MAX.
STANDBY CURRENT DRAIN .... 10 mA MAX.
OUTPUT CURRENT .... 500 mA MAX. (TO-5 PACKAGE)
1.5 mA MAX. (TO-3 PACKAGE)

LM110 VOLTAGE FOLLOWER

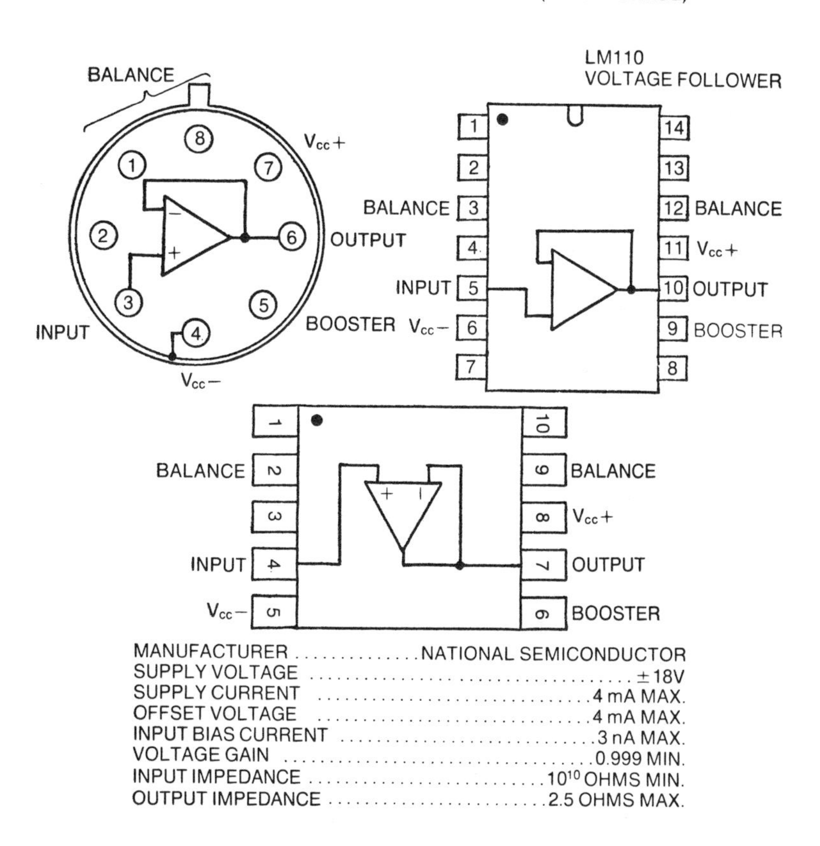

MANUFACTURER .... NATIONAL SEMICONDUCTOR
SUPPLY VOLTAGE .... ±18V
SUPPLY CURRENT .... 4 mA MAX.
OFFSET VOLTAGE .... 4 mA MAX.
INPUT BIAS CURRENT .... 3 nA MAX.
VOLTAGE GAIN .... 0.999 MIN.
INPUT IMPEDANCE .... 10^{10} OHMS MIN.
OUTPUT IMPEDANCE .... 2.5 OHMS MAX.

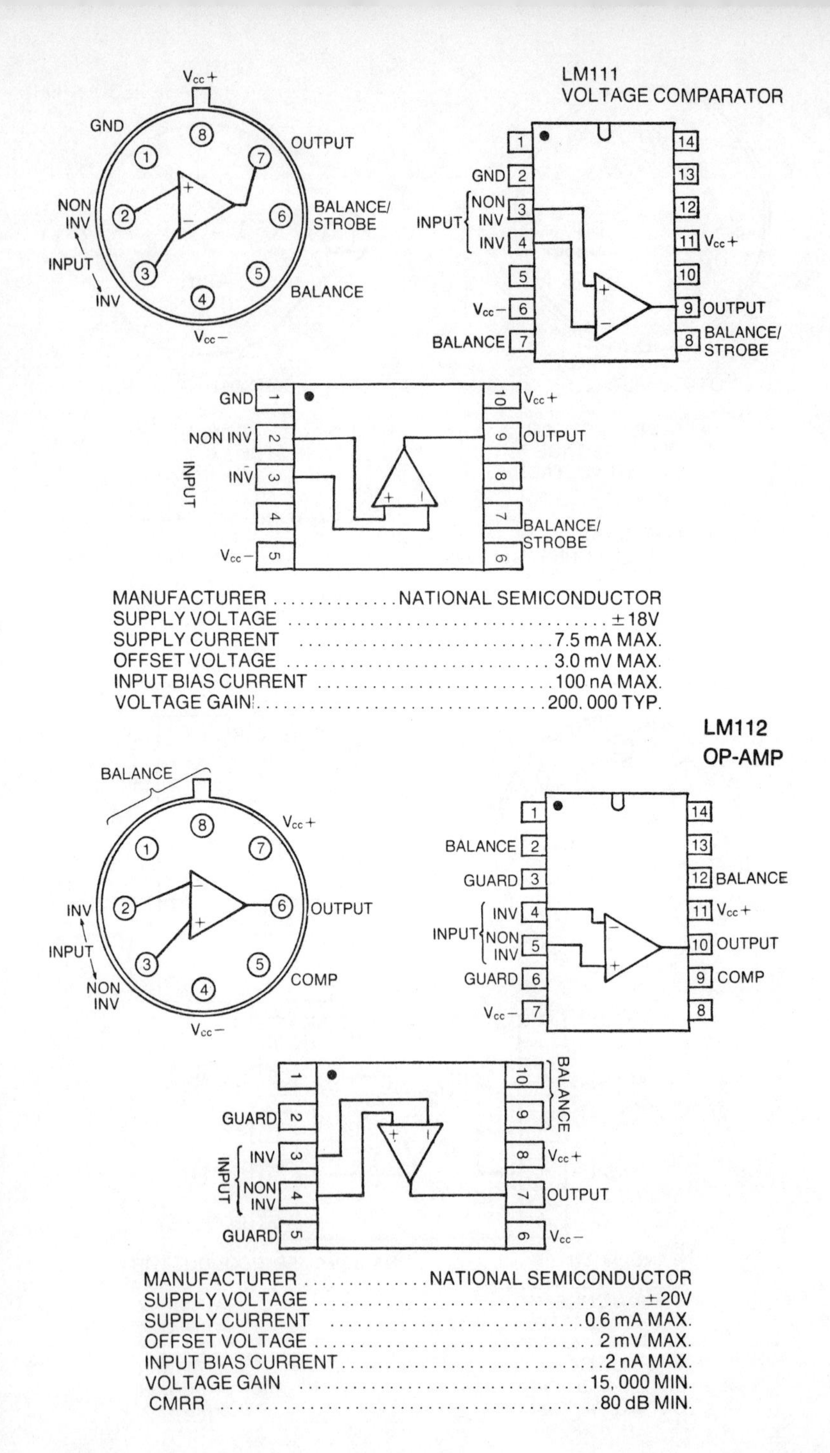

MANUFACTURER .... NATIONAL SEMICONDUCTOR
SUPPLY VOLTAGE .... ±18V
SUPPLY CURRENT .... 7.5 mA MAX.
OFFSET VOLTAGE .... 3.0 mV MAX.
INPUT BIAS CURRENT .... 100 nA MAX.
VOLTAGE GAIN .... 200, 000 TYP.

MANUFACTURER .... NATIONAL SEMICONDUCTOR
SUPPLY VOLTAGE .... ±20V
SUPPLY CURRENT .... 0.6 mA MAX.
OFFSET VOLTAGE .... 2 mV MAX.
INPUT BIAS CURRENT .... 2 nA MAX.
VOLTAGE GAIN .... 15, 000 MIN.
CMRR .... 80 dB MIN.

LM118
OP-AMP

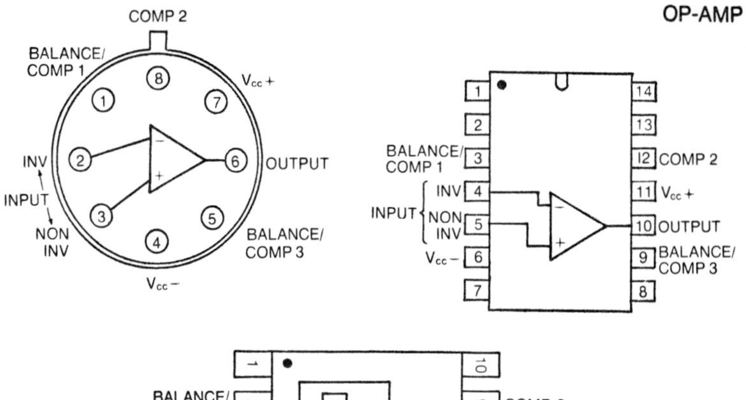

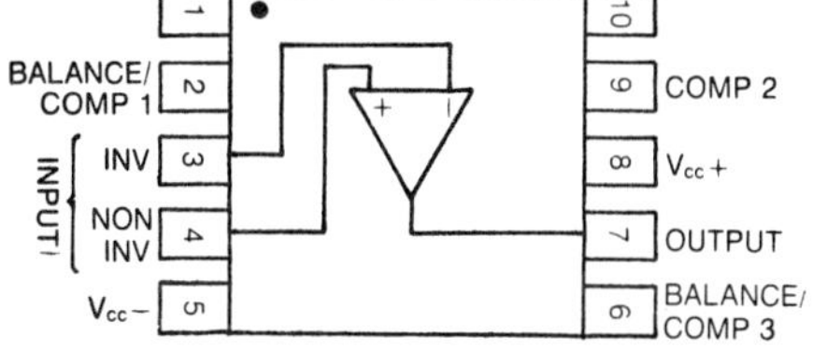

MANUFACTURER	NATIONAL SEMICONDUCTOR
SUPPLY VOLTAGE	±20V.
SUPPLY CURRENT	8 mA MAX.
OFFSET VOLTAGE	4 mV MAX.
INPUT BIAS CURRENT	250 nA MAX.
VOLTAGE GAIN	25,000 MIN.
CMRR	80 dB MIN.
BANDWIDTH	15 MHz
SLEW RATE	50V/μs MIN.

LM139
FOUR-CHANNEL COMPARATOR

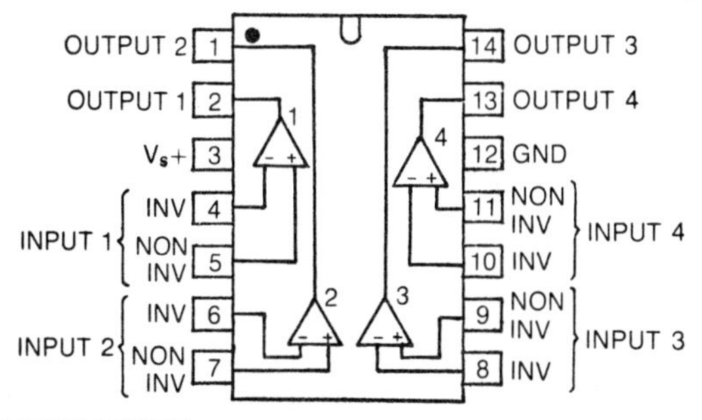

MANUFACTURER	NATIONAL SEMICONDUCTOR
SUPPLY VOLTAGE	36 V OR ±18V
SUPPLY CURRENT	2 mA MAX.
OFFSET VOLTAGE	5 mV MAX
INPUT BIAS CURRENT	25 nA MAX.
POWER DISSIPATION	900 mW (DIP) 800 mW (FLAT PACK)

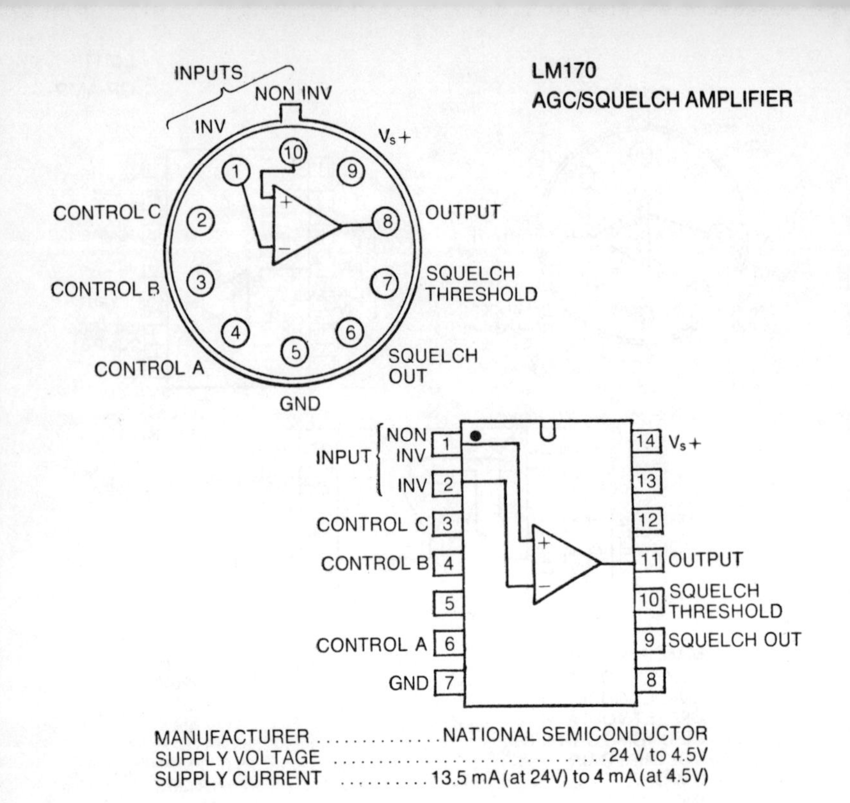

MANUFACTURER .... NATIONAL SEMICONDUCTOR
SUPPLY VOLTAGE .... 24 V to 4.5V
SUPPLY CURRENT .... 13.5 mA (at 24V) to 4 mA (at 4.5V)

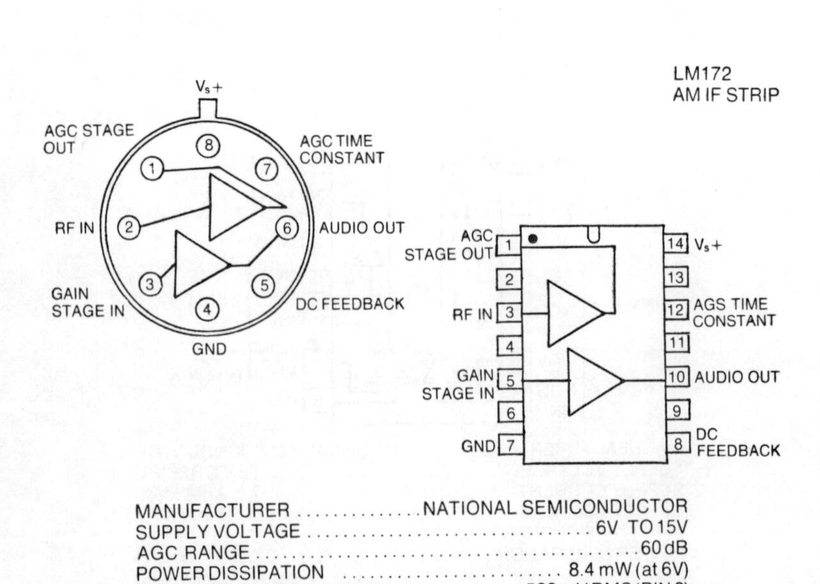

MANUFACTURER .... NATIONAL SEMICONDUCTOR
SUPPLY VOLTAGE .... 6V TO 15V
AGC RANGE .... 60 dB
POWER DISSIPATION .... 8.4 mW (at 6V)
REQUIRED RF INPUT LEVEL .... 500 mV RMS (PIN 2)

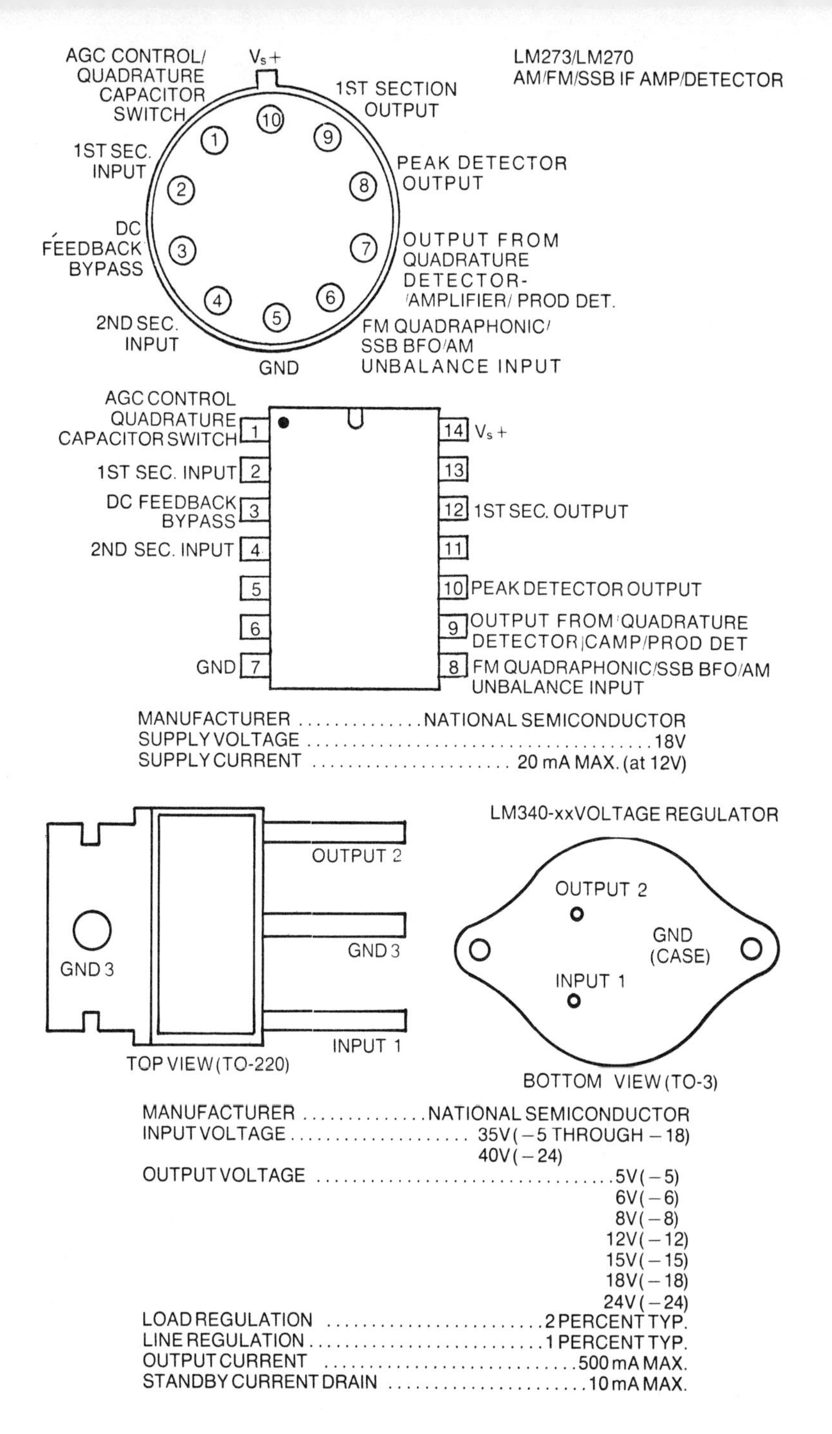
LM273/LM270
AM/FM/SSB IF AMP/DETECTOR
AGC CONTROL/
QUADRATURE
CAPACITOR
SWITCH
Vs+
1ST SECTION
OUTPUT
1ST SEC.
INPUT
PEAK DETECTOR
OUTPUT
DC
FEEDBACK
BYPASS
OUTPUT FROM
QUADRATURE
DETECTOR-
/AMPLIFIER/ PROD DET.
2ND SEC.
INPUT
FM QUADRAPHONIC/
SSB BFO/AM
UNBALANCE INPUT
GND
1
2
3
4
5
6
7
8
9
10
AGC CONTROL
QUADRATURE
CAPACITOR SWITCH
1ST SEC. INPUT
DC FEEDBACK
BYPASS
2ND SEC. INPUT
GND
14 Vs+
13
12 1ST SEC. OUTPUT
11
10 PEAK DETECTOR OUTPUT
9 OUTPUT FROM QUADRATURE
DETECTOR|CAMP/PROD DET
8 FM QUADRAPHONIC/SSB BFO/AM
UNBALANCE INPUT
MANUFACTURER ....NATIONAL SEMICONDUCTOR
SUPPLY VOLTAGE 18V
SUPPLY CURRENT .... 20 mA MAX. (at 12V)
LM340-xxVOLTAGE REGULATOR
OUTPUT 2
GND 3
GND 3
INPUT 1
TOP VIEW (TO-220)
OUTPUT 2
GND
(CASE)
INPUT 1
BOTTOM VIEW (TO-3)
MANUFACTURER ....NATIONAL SEMICONDUCTOR
INPUT VOLTAGE .... 35V (−5 THROUGH −18)
40V (−24)
OUTPUT VOLTAGE ....5V (−5)
6V (−6)
8V (−8)
12V (−12)
15V (−15)
18V (−18)
24V (−24)
LOAD REGULATION ....2 PERCENT TYP.
LINE REGULATION ....1 PERCENT TYP.
OUTPUT CURRENT ....500 mA MAX.
STANDBY CURRENT DRAIN ....10 mA MAX.

LM380
AUDIO POWER AMPLIFIER

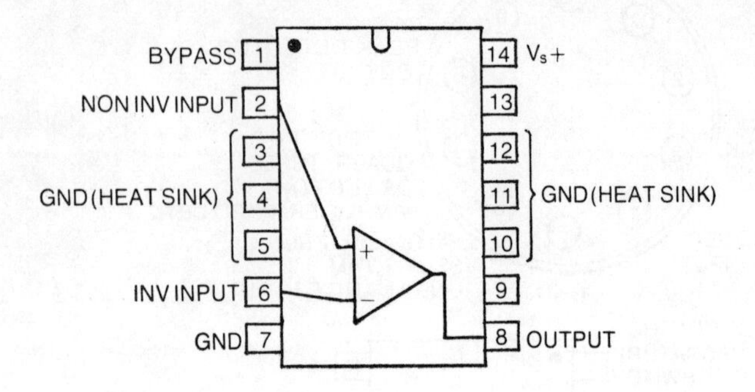

MANUFACTURER .... NATIONAL SEMICONDUCTOR
SUPPLY VOLTAGE .... 22V (8V MIN.)
SUPPLY CURRENT .... 1.3A MAX (OUTPUT SHORT-CIRCUITED)
OUTPUT POWER .... 2.5 RMS (.8 OHMS LOAD)
VOLTAGE GAIN .... 40 MIN.
INPUT IMPEDANCE .... 150K TYP.
TOTAL HARMONIC DISTORTION .... 0.2 PERCENT
STANDBY CURRENT DRAIN .... 25 mA MAX.
BANDWIDTH .... 100 kHz

LM381
LOW NOISE DUAL PREAMPLIFIER

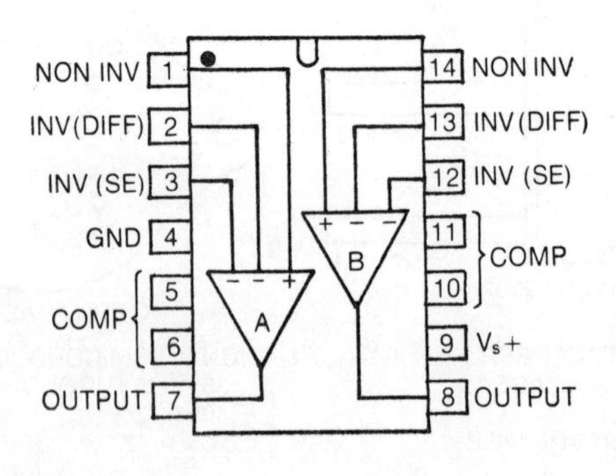

MANUFACTURER .... NATIONAL SEMICONDUCTOR
SUPPLY VOLTAGE .... 40V (9V MIN.)
SUPPLY CURRENT .... 10 mA (9 to 40V)
POWER DISSIPATION .... 800 mW
VOLTAGE GAIN .... 160,000 TYP.
TOTAL HARMONIC DISTORTION .... 0.1 PERCENT
NOISE FIGURE .... 1.6 dB MAX.

709
OP-AMP

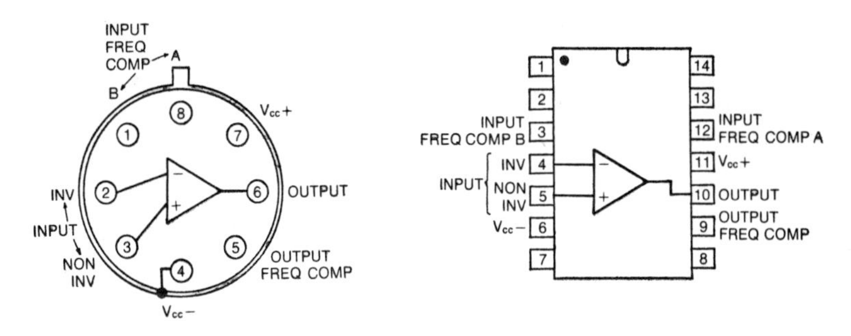

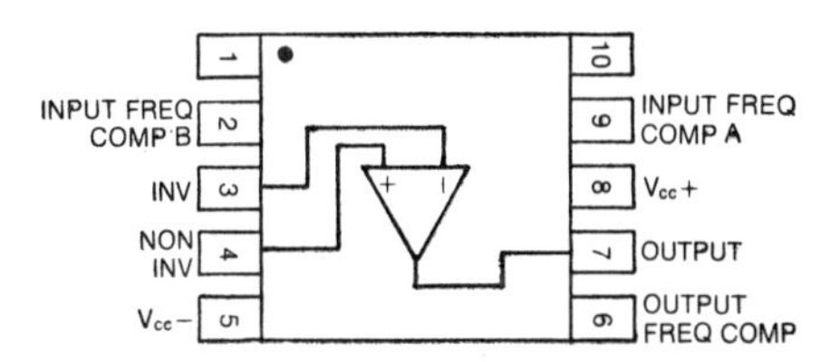

MANUFACTURER	VARIOUS
SUPPLY VOLTAGE	±18V
SUPPLY CURRENT	5.5 mA MAX.
OFFSET VOLTAGE	6 mV MAX.
INPUT BIAS CURRENT	0.5 μA MAX.
VOLTAGE GAIN	45, 000 TYP.
CMRR	70 dB MIN.

723
VOLTAGE REGULATOR

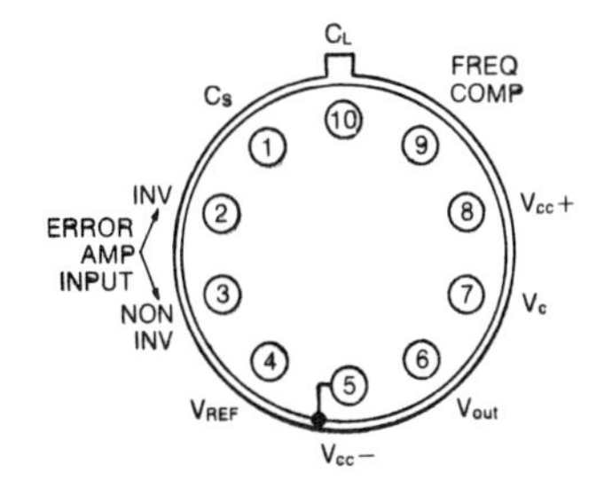

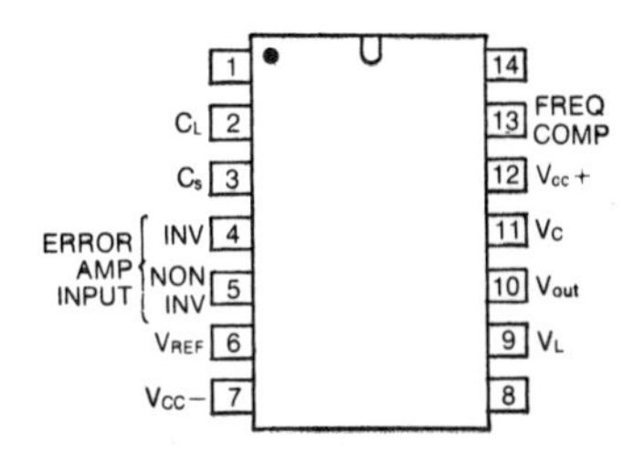

MANUFACTURER	VARIOUS
INPUT VOLTAGE RANGE	9.5 to 40V
OUTPUT VOLTAGE RANGE	2.0 to 37V
INPUT-OUTPUT DIFFERENTIAL	5V (FAIRCHILD) UP TO 40 V (NATIONAL SEMICONDUCTOR)
LOAD REGULATION	0.2 PERCENT MAX.
LINE REGULATION	0.5 PERCENT MAX.
STANDBY CURRENT DRAIN	4 mA MAX.

741
OF OP-AMP

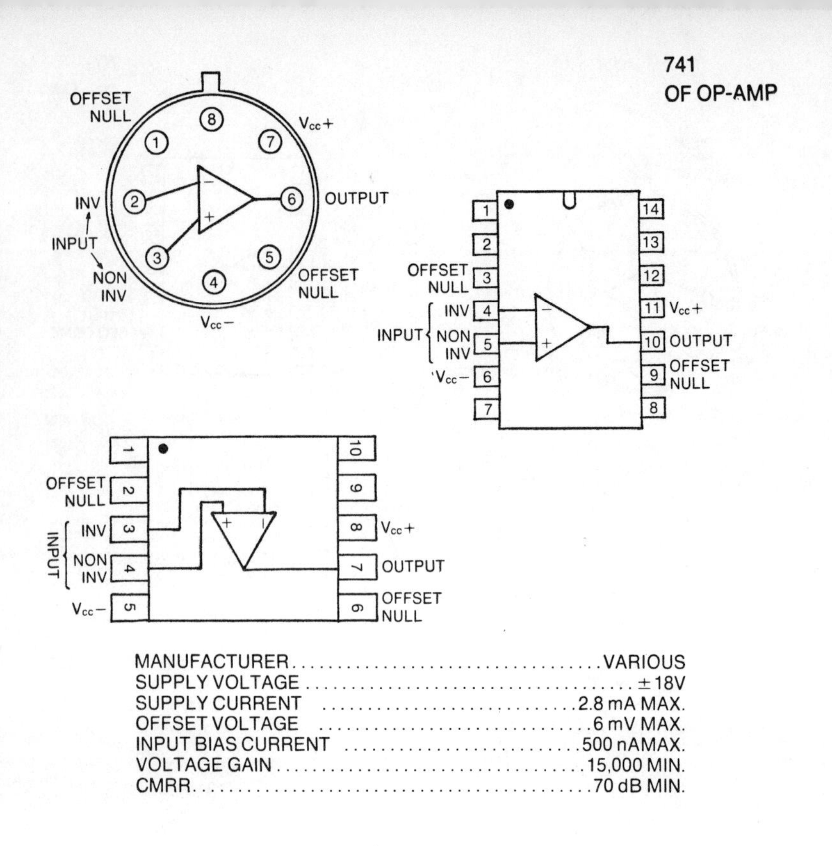

MANUFACTURER	VARIOUS
SUPPLY VOLTAGE	±18V
SUPPLY CURRENT	2.8 mA MAX.
OFFSET VOLTAGE	6 mV MAX.
INPUT BIAS CURRENT	500 nAMAX.
VOLTAGE GAIN	15,000 MIN.
CMRR	70 dB MIN.

747
DUAL OP-AMP

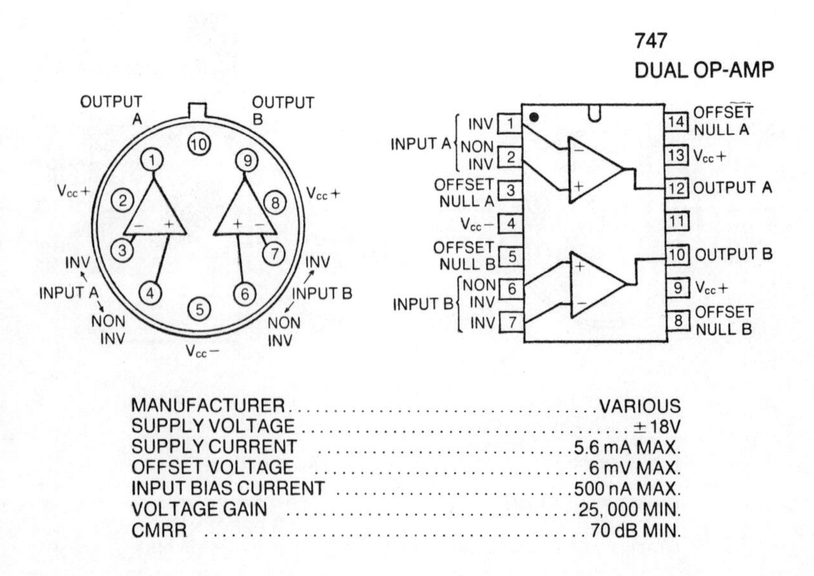

MANUFACTURER	VARIOUS
SUPPLY VOLTAGE	±18V
SUPPLY CURRENT	5.6 mA MAX.
OFFSET VOLTAGE	6 mV MAX.
INPUT BIAS CURRENT	500 nA MAX.
VOLTAGE GAIN	25, 000 MIN.
CMRR	70 dB MIN.

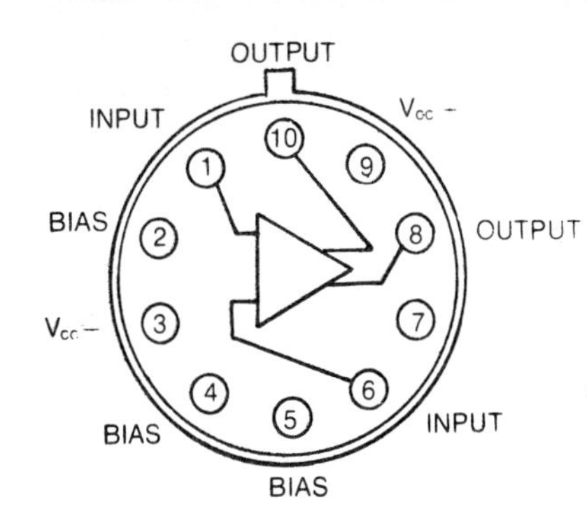

CA3000
DIFFERENCE AMPLIFIER

MANUFACTURER	RCA
SUPPLY VOLTAGE	±6V
OFFSET VOLTAGE	5 mV MAX.
INPUT BIAS CURRENT	36 μA MAX.
VOLTAGE GAIN	25 MIN.
CMRR	70 dB MIN.

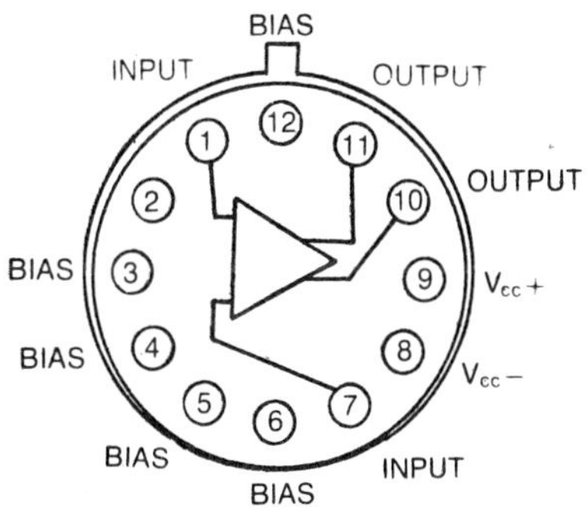

CA3005/CA3006
RF DIFFERENCE AMPLIFIER

MANUFACTURER	RCA
SUPPLY VOLTAGE	±6V
OFFSET VOLTAGE	5 mV MAX (CA3005) 1 mV MAX (CA3006)
INPUT BIAS CURRENT	40μA MAX.
POWER GAIN	16 dB MIN. (at 100 MHz)
CMRR	101 dB TYP.
NOISE FIGURE	9 dB MAX. (at 100 MHz)

CA3015
OP-AMP

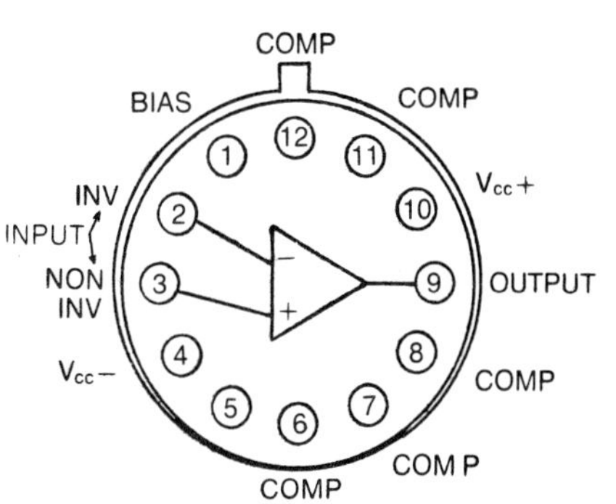

MANUFACTURER	RCA
SUPPLY VOLTAGE	±12V
OFFSET VOLTAGE	5 mV MAX.
INPUT BIAS CURRENT	24 μA MAX.
VOLTAGE GAIN	2000 MIN.
CMRR	80 dB MIN.

CA3019
DIODE ARRAY

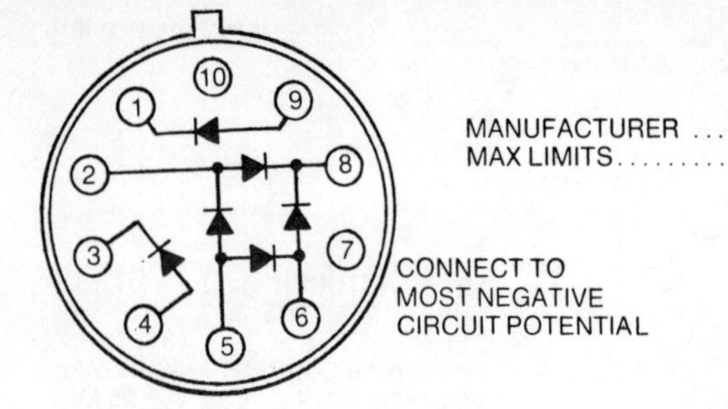

MANUFACTURER .... RCA
MAX LIMITS .... +12 V, −3V ON EACH PIN
WITH PIN 7 AT −6V DC

CA3020/CA3020A
WIDE BAND POWER AMPLIFIER

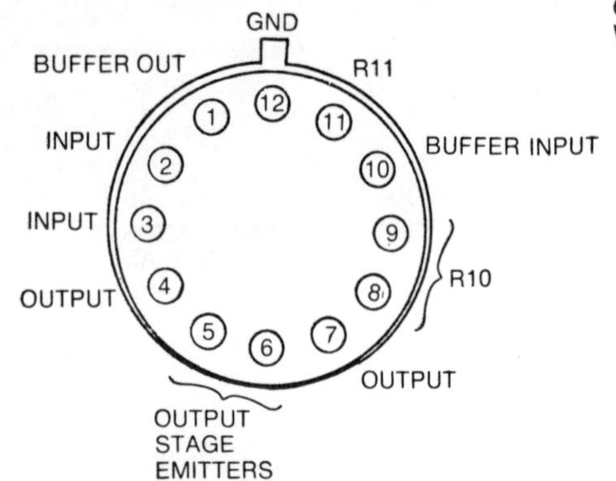

MANUFACTURER .... RCA
SUPPLY VOLTAGE .... 3 TO 9V (CA3020)
3 TO 12V (CA3020A)
SUPPLY CURRENT .... 35 mA MAX. (CA3020)
30 mA MAX. (CA3020A)
POWER OUTPUT .... 550 mW RMS (CA3020)
1W RMS (CA3020A)
POWER GAIN .... 75 dB TYP.
INPUT IMPEDANCE .... 55K TYP.
SIGNAL TO NOISE RATION .... 70 dB TYP. (CA3020)
66 dB TYP. (CA3020A)
TOTAL HARMONIC DISTORTION ...... 3.1 PERCENT (CA3020)
3.3 PERCENT (CA3020A)

CA3028
DIFFERENCE/CASCODE AMPLIFIER

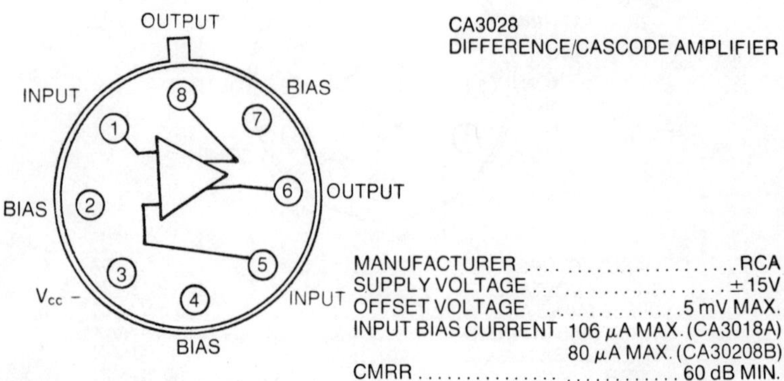

MANUFACTURER .... RCA
SUPPLY VOLTAGE .... ±15V
OFFSET VOLTAGE .... 5 mV MAX.
INPUT BIAS CURRENT 106 μA MAX. (CA3018A)
80 μA MAX. (CA30208B)
CMRR .... 60 dB MIN.

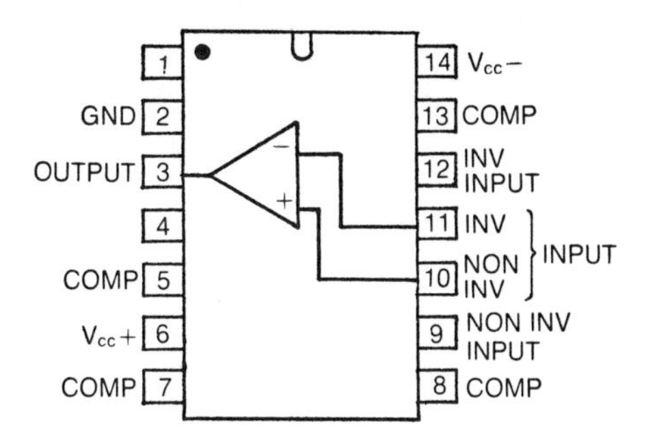

MANUFACTURER .... RCA
SUPPLY VOLTAGE .... ±12V (CA3033/CA3047)
±15V (CA3033A/CA3047A)
INPUT OFFSET CURRENT ...... 35 nA MAX. (CA3033/CA3047A)
25 nA MAX. (CA3033A/CA3047A)
INPUT BIAS CURRENT .... 350 nA MAX. (CA3033/CA3047)
180 nA MAX. (CA3033A/CA3047A)
VOLTAGE GAIN .... 16,000 TYP. (CA3033/CA3047)
22,000 TYP. (CA3033A/CA3047A)
CMRR .... 84 dB MIN. (CA3033/CA3047)
93 dB MIN. (CA3033A/CA3033A/CA3047A)
POWER OUTPUT .... 80 mW MIN. (CA3033A3047)
(500 OHMS LOAD) .... 220 mW MIN. (CA3033A/CA/CA3047A)

CA3048
QUAD AC AMPLIFIER ARRAY

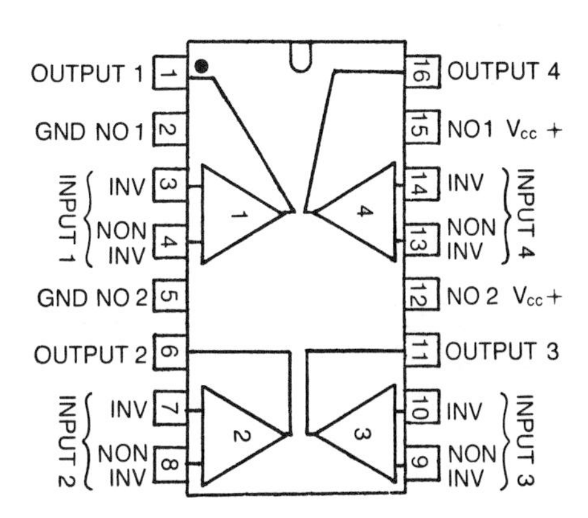

MANUFACTURER .... RCA
SUPPLY VOLTAGE 16V
INPUT RESISTANCE .... 90K TYP.
OUTPUT RESISTANCE .... 1000 OHMS TYP.
VOLTAGE GAIN .... 450 MIN.
NOISE FIGURE .... 2 dB TYPE (at 1 kHz)
BANDWIDTH .... 300 kHz TYP.

CA3062
PHOTO DETECTOR AND POWER AMPLIFIER

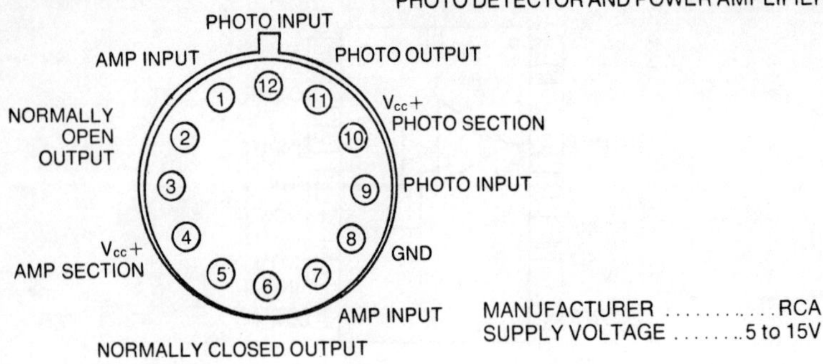

MANUFACTURER ....RCA
SUPPLY VOLTAGE ....5 to 15V

CA3080
OPERATIONAL TRANSCONDUCTANCE AMPLIFIER

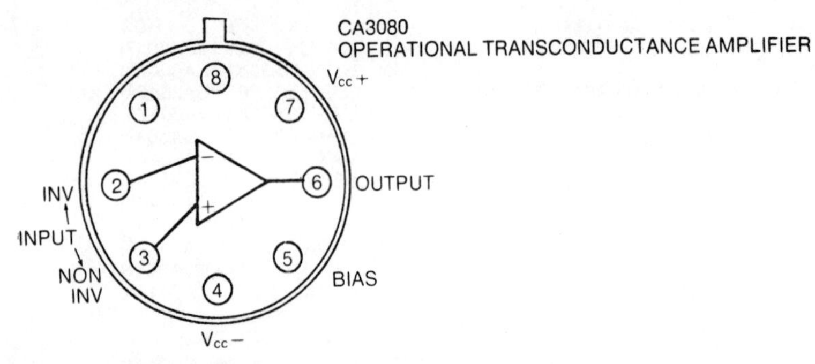

MANUFACTURER ....RCA
SUPPLY VOLTAGE ....±18V
SUPPLY CURRENT ....1.2 mA MAX.
OFFSET VOLTAGE ....5 mV MAX.
INPUT BIAS CURRENT ....5 μ MAX.
FORWARD TRANSCONDUCTANCE ....6700 μmho MIN.
CMRR ....80 dB MIN.
BANDWIDTH ....2 MHz

CA3081
COMMON EMITTER TRANSISTOR ARRAY

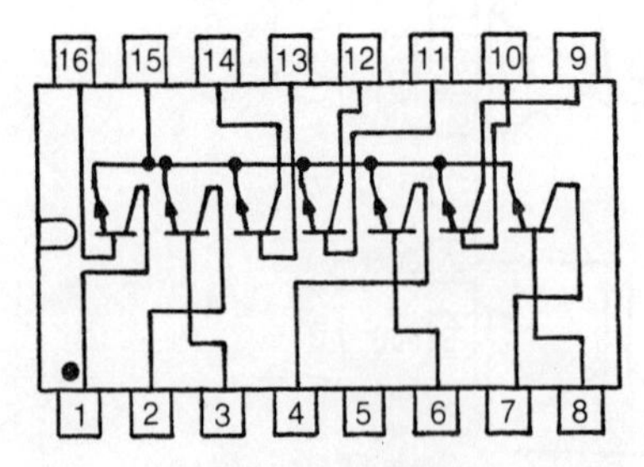

MANUFACTURER ....RCA
COLLECTOR-BASE BREAKDOWN ......20V MIN ($I_c = 500\ \mu A$)
COLLECTOR-SUBSTRATE BREAKDOWN ....20V MIN.
COLLECTOR-EMITTER BREAKDOWN ....16V MIN.
EMITTER-BASE BREAKDOWN ....5V MIN.
h_{FE} ....30 MIN.
V_{BE} SAT ....0.87 TYP.
I_{CEO} ....10 μA
I_{CBO} ....1 μA

CA3082
COMMON COLLECTOR TRANSISTOR ARRAY

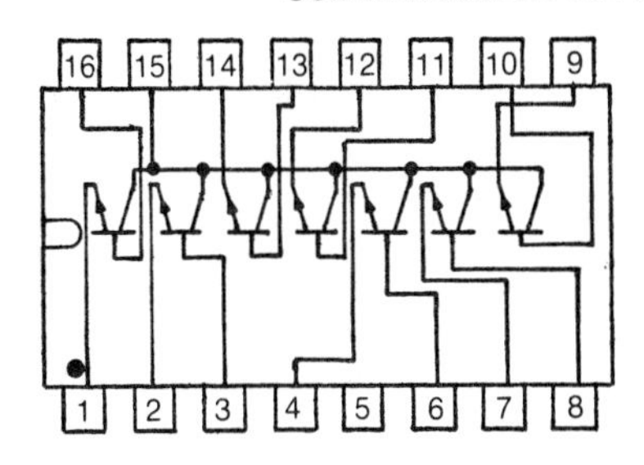

MANUFACTURER .... RCA
COLLECTOR-BASE BREAKDOWN ..... 20V MIN. (I_c = 500 μA)
COLLECTOR-SUBSTRATE BREAKDOWN .... 20V MIN.
COLLECTOR-EMITTER BREAKDOWN .... 16V MIN.
EMITTER-BASE BREAKDOWN .... 5V MIN.
h_{FE} .... 30 MIN.
V_{BE} SAT .... 0.87 TYP.
I_{CEO} .... 10 μA
I_{CBO} .... 1 μA

CA3085
POSITIVE VOLTAGE REGULATOR

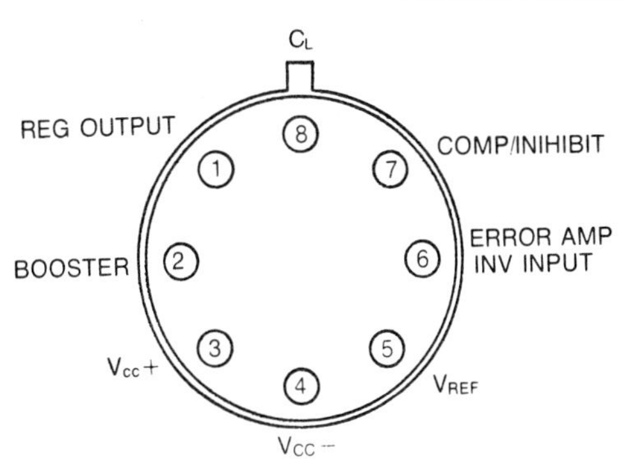

MANUFACTURER .... RCA
INPUT VOLTAGE RANGE .... 7.5 to 30V (CA3085)
7.5 to 40V (CA3085A)
7.5 to 50V (CA3085B)
OUTPUT VOLTAGE RANGE ..... 1.6 to 27 TYP. (CA3085)
1.6 to 37 TYP. (CA3085A)
1.6 to 47 TYP. (CA3085B)
INPUT-OUTPUT DIFFERENTIAL .. 4 to 28V MAX. (CA3085)
4 to 38V (CA3085A)
4 to 48V MAX. (CA3085B)
LOAD REGULATION .... 0.15 PERCENT MAX.
LINE REGULATION .... 0.15 PERCENT MAX. (CA3085)
0.10 PERCENT MAX. (CA3085A)
0.08 PERCENT (CA3085B)
STANDBY CURRENT DRAIN ..... 4.5 mA MAX. (CA3085)
5 mA MAX. (CA3085A)
7 mA MAX. (CA3085B)

CA3094
PROGRAMMABLE POWER SWITCH AMPLIFIER

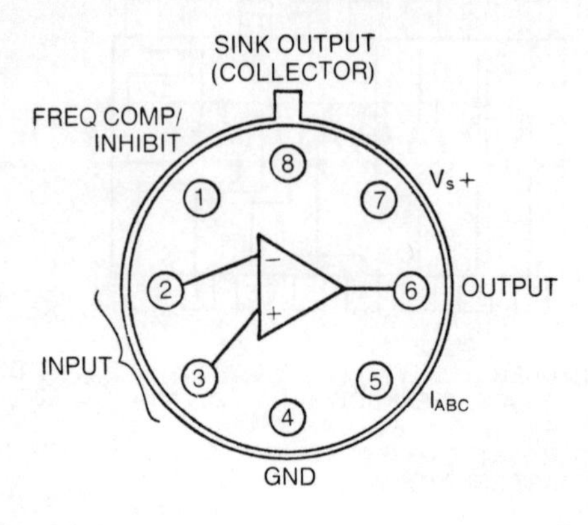

MANUFACTURER ....RCA
SUPPLY VOLTAGE ....±12V or +24V (CA3094T)
±18V or +36V (CA3094AT)
±22V or +44V (CA3094BT)
OFFSET VOLTAGE ....5 mV MAX.
INPUT BIAS CURRENT ....500 nA MAX.
VOLTAGE GAIN ....20,000 MIN.
CMRR ....70 dB MIN.
BANDWIDTH ....30 MHz
POWER DISSIPATION..........630 mW (WITHOUT HEAT SINK)
1.6 W (WITH HEAT SINK)

LM3900
QUAD "NORTON" AMPLIFIER

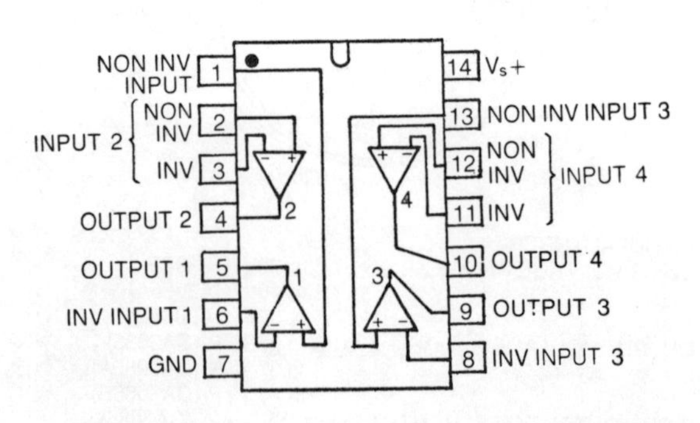

MANUFACTURER ....NATIONAL SEMICONDUCTOR
SUPPLY VOLTAGE ....±18V OR +36V
SUPPLY CURRENT ....10 mA MAX.
INPUT BIAS CURRENT ....200 nA MAX.
VOLTAGE GAIN ....1200 MIN. (at 100 Hz)
BANDWIDTH ....2.5 MHz

NOTE: DEVICE IS DESIGNED TO OPERATE FROM SUPPLY VOLTAGES AS LOW AS ±2 OR +4V.

LM4250
PROGRAMMABLE OP-AMP

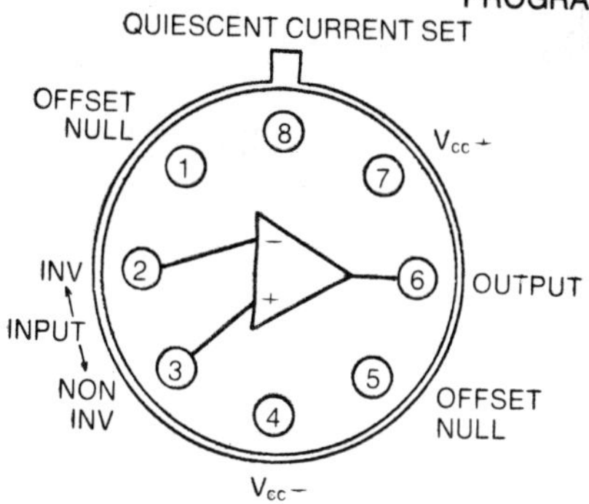

MANUFACTURER . . . . NATIONAL SEMICONDUCTOR
SUPPLY VOLTAGE . . . . ± 1V TO ± 18V
SUPPLY CURRENT 90 μ MAX.(DEPENDING ON CURRENT SET)
OFFSET VOLTAGE . . . . 5 mV MAX.
INPUT BIAS CURRENT . . . . 50 nA MAX.
VOLTAGE GAIN . . . . 30, 000 MIN.
CMRR . . . . 70 dB MIN.

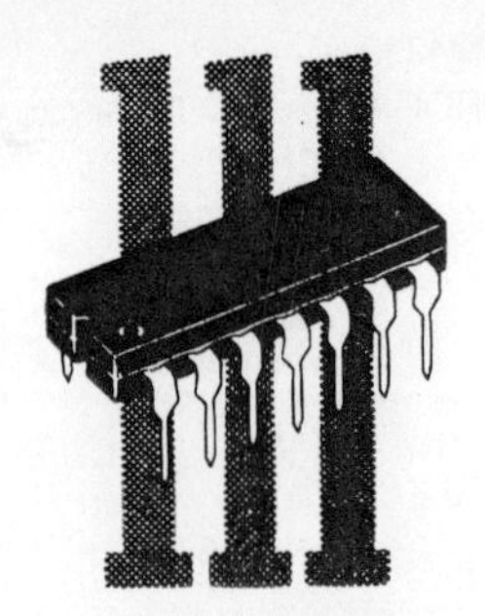

Glossary

This list, while by no means exhaustive, explains many of the more unfamiliar terms associated with linear ICs.

active filter—A high-pass, low-pass, bandpass, or band-elimination filter that uses an active element, such as an operational amplifier and relatively small capacitors, rather than larger inductors and capacitors that would be required in a conventional passive filter.

common-mode gain (CMG)—A characteristic of amplifiers with a differential input. Common-mode gain refers to the voltage gain of a common-mode signal. Typical values in op-amps are around minus 30 dB.

common-mode rejection ration (CMRR)—The difference in dB between common-mode gain and differential-mode gain in an op-amp.

common-mode signal—In an amplifier with a differential input, a signal, referred to ground, that appears at both inverting and noninverting inputs with the same phase, amplitude, and frequency. Power-line hum is the most frequently encountered common-mode signal.

comparator—An active device that provides a logical zero when its input is below a preset reference value and a logical one when its input is above that value.

difference amplifier—The basic input stage of most operational amplifiers. An amplifier with an inverting and a noninverting input. The output voltage is a function of the voltage difference between the two inputs.

differential-mode gain (DMG)—The voltage gain exhibited by an operational amplifier in response to differential-mode signals.

differential-mode signal—In an amplifier with a differential input, a signal that appears at inverting and noninverting inputs with opposite phase, but identical frequency and amplitude. It is not necessarily referred to ground

feedback—The process of coupling some of the output of an amplifier back to its input. Negative feedback reduces the gain of an amplifier, but has compensating beneficial results. Positive feedback can be used to boost gain (regeneration), but usually results nin oscillation.

feed-forward—A frequency-compensation technique in operational amplifiers. A small-value capacitor is used to bypass a gain stage that has poor performance at high frequency.

foldback—A technique for protecting voltage regulators from short circuits. After a certain output-current level is reached, any further load on the regulator results in less, rather than more current flow.

guarding—A method of protecting the inputs to a high-gain op-amp by surrounding the input terminals with a conducting ring of printed circuit board conductor. This isolates the inputs from potential leakage currents from other parts of the circuit.

hybrid integrated circuit—An IC consisting of one or more monolithic IC chips together with discrete passive components. Hybrid ICs, when they become commercially available, will offer more circuit complexity than can be achieved with present generation monolithic ICs.

LSI—Short for *large scale integration*, the technology that produces microcircuits with 100 or more active devices on a single chip. Functional blocks that include several op-amps and other devices are examples of LSI devices.

level shifting—The process of changing a differential signal input to a single-ended output within an operational amplifier.

MSI—Short for *medium-scale integration*, the technology that produces microcircuits with 15 or more active devices on a single chip. Most op-amps are MSI devices.

monolithic IC—An integrated circuit in which all components are created by *pn* junctions grown on a single chip of silicon crystal.

offset—The voltage that must be applied between the inverting and noninverting inputs of an operational amplifier to cause the output voltage to go to zero.

operational amplifier (op-amp)—An active device characterized by high input impedance and very high voltage gain. Operational amplifiers generally have a differential input and a single-ended output.

operational transconductance amplifier (OTA)—A device similar to a conventional operational amplifier, but with a current output and a means for controlling tranconductance with a current input.

phase-locked loop (PLL)—A closed-loop system, consisting of a phase detector, a filter, and a voltage-controlled oscillator. The phase detector provides an error signal that locks the voltage-controlled oscillator to the frequency of an incoming signal.

pullup—A DC voltage imposed on the input of an amplifier to move the amplifier's operating point out of the offset range. Pullup is usually accomplished by means of a voltage divider network.

settling time—The time necessary for an operational amplifier output to slew through a defined voltage change and settle to within a defined error of its final voltage.

slew rate—The rate of change of an op-amp's output voltage, in volts per microsecond, in response to a step change in input voltage.

single-ended signal—As opposed to a difference-mode signal, a signal that is at ground potential when it is at zero level.

Index

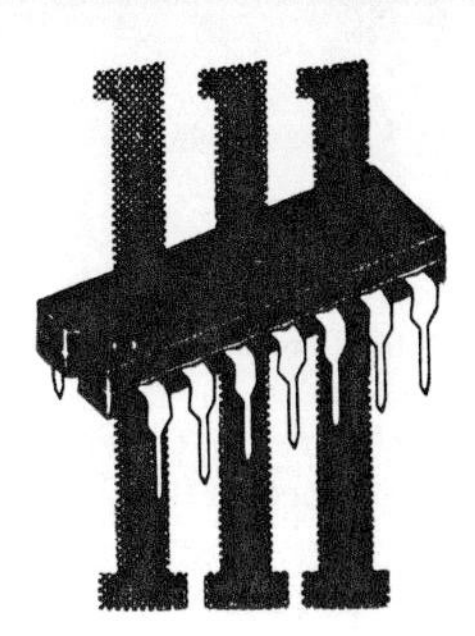

IC Index